高等学校土建类学科专业"十四五"系列教材
高等学校土木工程专业应用型本科系列教材

土木建筑类专业英语

Professional English for Civil and Architectural Engineering

占丰林　蔡丽蓉　主　编
易秀娟　副主编

中国建筑工业出版社
CHINA ARCHITECTURE & BUILDING PRESS

图书在版编目(CIP)数据

土木建筑类专业英语=Professional English for Civil and Architectural Engineering／占丰林，蔡丽蓉主编；易秀娟副主编. —北京：中国建筑工业出版社，2022.12

高等学校土建类学科专业"十四五"系列教材　高等学校土木工程专业应用型本科系列教材

ISBN 978-7-112-28141-1

Ⅰ. ①土…　Ⅱ. ①占…②蔡…③易…　Ⅲ. ①土木工程–英语–高等学校–教材　Ⅳ. ①TU

中国版本图书馆 CIP 数据核字(2022)第 209387 号

责任编辑：赵　莉　吉万旺
文字编辑：卜　煜
责任校对：张惠雯

高等学校土建类学科专业"十四五"系列教材
高等学校土木工程专业应用型本科系列教材
土木建筑类专业英语
Professional English for Civil and Architectural Engineering
占丰林　蔡丽蓉　主　编
易秀娟　副主编

*

中国建筑工业出版社出版、发行(北京海淀三里河路9号)
各地新华书店、建筑书店经销
北京科地亚盟排版公司制版
北京云浩印刷有限责任公司印刷

*

开本：787 毫米×1092 毫米　1/16　印张：17　字数：497 千字
2023 年 3 月第一版　　2023 年 3 月第一次印刷
定价：**60.00** 元（赠教师课件）
ISBN 978-7-112-28141-1
(40293)

版权所有　翻印必究
如有印装质量问题，可寄本社图书出版中心退换
（邮政编码　100037）

内 容 提 要

全书共分 29 个单元，主要内容包括：笛卡尔坐标与解析几何，函数及其导数，计算机硬件，办公自动化，应力、应变与胡克定律，梁的挠曲线微分方程及弯矩面积法，圆杆扭转与剪切胡克定律，临界荷载与欧拉公式，土木与建筑工程，"建筑经济与管理"课程思政探索，土木工程材料，土力学与基础，砌体结构，钢筋混凝土与预应力混凝土，混凝土：支模、扎筋、浇捣和养护，建筑结构体系，钢结构连接与施工，高速公路，斜拉桥，隧道工程，中国皇城规划特征，城市总体规划步骤与考虑因素，城市交通规划阶段，住房设计准则与住宅规划考虑因素，包豪斯——"艺术与技术的新统一"，护德艺术博物馆，华南理工大学-都灵理工大学联队设计的长屋与教授的绿色建筑理念，3D 打印和注浆锚索的 ANSYS 分析、施工技术及锚固机理（胀锚拱理论）等涉及高等数学、计算机基础、工程力学、土木工程、隧道工程、交通工程、城乡规划和建筑学等方面的内容，特别是增加了裂隙岩体锚固机理及近几年兴起的课程思政内容，还遴选了两篇具有时代特色的教学研究前沿文章——绿色建筑和 3D 打印。为便于教学和读者查阅，每个单元分为教学目标、课文、词汇与短语、注释和习题五个部分，最后一个单元还加了英文科技论文的写作技巧。书后附有常用数学符号公式的读法和英汉对照常用土木建筑类专业词汇表。

本书可作为大土木环境下高等院校土木建筑类专业（如土木工程、隧道工程、交通工程、城乡规划和建筑学专业等）高年级本科生、研究生开设"土木建筑类专业英语"课程所用的教材，也可作为土木建筑类专业教师、研究人员和具有一定英语基础的工程技术人员及自学者学习参考，特别是作为土木建筑类专业涉外施工管理课程和涉外施工人员英语培训或自学教材，也可作为准备出国做访问学者或攻读学位人员的复习用书。

为便于课堂教学，本书配有教学课件，请选用此教材的教师通过以下方式获取课件：邮箱：jckj@cabp.com.cn；电话：(010) 58337285；建工书院：http://edu.cabplink.com。

前言

在大土木环境下，许多课程的设置和内容都进行了调整。目前我国土木建筑行业与国外交流甚多，并参与国际市场的竞争，许多专业技术人员希望有一本全面系统的土木建筑类专业英语参考书，涉及土木建筑类主要公共基础课程（如高等数学、计算机基础）、主要专业基础课程（如工程力学）和多学科多专业的主要专业课程（如土木工程、隧道工程、交通工程、城乡规划、建筑学）方面的英文文献和常用专业词汇，本书就是专为此目标编写的教材。一般某个专业的专业英语教材只是将本专业的英文文献和常用专业词汇编写进去，目前市场上只有数学专业英语、计算机专业英语、工程力学专业英语、土木工程专业英语、隧道工程专业英语、交通工程专业英语、城乡规划专业英语和建筑学专业英语等适用于单个专业的专业英语教材，而在大土木环境下学者和工程技术人员在全英文教学、双语教学、撰写英文科技论文和涉外交流（尤其是撰写科技论文和口头交流方面）时需要的知识面非常广，涉及大土木的多学科多专业的知识以及运用高等数学、工程力学和计算机软件模拟分析等内容。出差在外携带八九本书籍实在不方便，因此，大土木环境下许多学者和专业技术人员都希望能有一本全面系统的土木建筑类专业英语参考书，以达到"一册在手全都有，通聊天下无敌手！"之效果。

本教材不仅将大土木的土木工程、隧道工程、交通工程、城乡规划、建筑学等多个专业的专业英文文献和常用专业词汇编入，还将土木建筑类专业主要公共基础课程（如高等数学、计算机基础）和主要专业基础课程（如工程力学）也编入进去，这在专业英语教材的编写历史上是独一无二的，是融合创新。为此，本书的编写尽可能做到内容的系统性、知识性和实用性，使本书除可用作大土木环境下高等院校土木建筑类专业（如土木工程、隧道工程、交通工程、城乡规划和建筑学专业等）高年级本科生、研究生开设"土木建筑类专业英语"课程所用的教材外，也可作为土木建筑类专业的教师、研究人员和具有一定英语基础的工程技术人员及自学者学习参考，特别是作为土木建筑类专业涉外施工管理课程和涉外施工人员英语培训或自学教材，也可作为准备出国做访问学者或攻读学位人员的复习用书。

本书语言规范，合乎时代要求，具有很强的代表性。具体包括笛卡尔坐标与解析几何，函数及其导数，计算机硬件，办公自动化，应力、应变与胡克定律，梁的挠曲线微分方程及弯矩面积法，圆杆扭转与剪切胡克定律，临界荷载与欧拉公式，土木与建筑工程，柱屈曲，课程思政，土木工程材料，土力学与基础，砌体结构，钢筋混凝土与预应力混凝土，混凝土施工，建筑结构体系，钢结构连接与施工，高速公路，斜拉桥，隧道工程，中国皇城规划，城市总体规划，城市交通规划，住宅设计准则，包豪斯，护德艺术博物馆，绿色建筑，3D 打印和裂隙岩体锚固机理等涉及高等数学、计算机基础、工程力学、土木工程、隧道工程、交通工程、城乡规划和建筑学等方面的内容，特别是增加了裂隙岩体锚固机理及近几年兴起的课程思政内容，还遴选了两篇具有时代特色的教学研究前沿文章——绿色建筑和 3D 打印，使本教材具有一定的前瞻性。

本书共有 29 个单元，每个单元分为教学目标、课文、词汇与短语、注释和习题五个部分，最后一个单元还加了英文科技论文的写作技巧。在习题中配有（词组、句子或段落）中英互译、用英语回答问题或单项选择题以帮助读者理解欣赏课文等形式多样的内容，以确保基础英语和专业英语的衔接和过渡，并提高读者的学习兴趣。书后还附有常用数学符号公式的读法（这主要是针对在专业英语、全英文教学和双语教学过程中或涉外交流中出现的常用符号和复杂数学公式难以通过口语念出来的问题而专门设置的）和英汉对照常用土木建筑类专业词汇表，词汇包括土木建筑类专业的公共基础课程（高等数学、计算机基础）、专业基础课程（工程力学）和专业课程

（土木工程、隧道工程、交通工程、城乡规划、建筑学）方面的常用专业词汇。

　　本书由江西理工大学建筑与设计学院专业教师编写。主编占丰林教授主讲过"采矿工程专业英语""土木工程专业英语""城乡规划专业英语"和"建筑学专业英语"等课程，全英文讲授过"计算机辅助建筑设计"课程，并主讲过"建筑结构选型""三维动画"和"计算机建筑渲染"等双语课程，跨专业词汇量大，具有多个专业的专业英语授课经验和多门课程的双语课程教学经验；蔡丽蓉副教授一直从事城乡规划和建筑学的教学、科研和管理工作，拥有丰富的专业教学、科研和管理工作经验；易秀娟讲师是国家注册规划师，在讲授城乡规划专业主干课程的同时，经常深入乡村一线进行乡村规划研究，城乡规划理论和实践经验造诣较深。本书第1~20单元、第29单元以及附录由占丰林编写；第24~28单元由蔡丽蓉编写；第21~23单元由易秀娟编写。全书由占丰林统一修订定稿，本书为江西理工大学教务处教材建设计划项目，由江西理工大学资助出版。

　　在本书编写过程中得到了许多文献、网站和微信公众号作者的支持和帮助，在此一并向他们表示感谢。由于编写时间紧迫和编者水平有限，书中难免会有不足之处，热忱欢迎使用本书的读者提出宝贵意见和建议。

2022年4月19日

于江西理工大学建筑与设计学院

Contents

1 Cartesian Coordinate System and Analytic Geometry

（笛卡尔坐标与解析几何）

教学目标：本单元的主要内容有笛卡尔坐标系、笛卡尔平面直角坐标轴、坐标原点、横坐标、纵坐标、象限、毕达哥拉斯定理(勾股定理)、解析几何与微积分的关系以及笛卡尔方程。教学重点是笛卡尔坐标系，难点是笛卡尔方程。通过本单元的学习，要求掌握如何使用一对数字表达平面几何上的一个点这种笛卡尔思想以及如何使用笛卡尔方程表达一个几何图形。

1.1 Cartesian Coordinate System

We have certain objects (polygonal regions, circular regions, parabolic segments, etc.) whose areas we wish to measure. If we hope to arrive at a treatment of area that will enable us to deal with many different kinds of objects, we must first find an effective way to describe these objects. The most primitive way of doing this is by drawing figures, as was done by the ancient Greeks. *A much better way was suggested by Rene Descartes* (1596 ~ 1650), *who introduced the subject of analytic geometry* (*also known as Cartesian geometry*). [1] Descartes' idea is to represent geometric points by numbers. The procedure for points in a plane is this:

Two perpendicular reference lines (called coordinate axes) are chosen, one horizontal (called the "x-axis"), the other vertical (the "y-axis"). Their point of intersection, denoted by O, is called the origin. On the x-axis , a convenient point is chosen to the right of O and its distance from O is called the unit distance. Vertical distances along the y-axis are usually measured with the same unit distance, although sometimes it is convenient to use a different scale on the y-axis. Now, each point in the plane (sometimes called the xy-plane) is assigned a pair of numbers, called its coordinates. These numbers tell us how to locate the point. Fig. 1-1 illustrates some examples. The point with coordinates (3, 2) lies three units to the right of the y-axis and two units above the x-axis. The number 3 is called the x-coordinate of the point and 2 is the y-coordinate of the point. Points to the left of the y-axis have a negative x-coordinate; those below the x-axis have a negative y-coordinate. The x-coordinate of a point is sometimes called its abscissa and the y-coordinate is called its ordinate.

When we write a pair of numbers such as (a, b) to represent a point, we agree that the abscissa or x-coordinate, a, is written first. For this reason, the pair (a, b) is often referred to as an ordered pair. It is clear that two ordered pairs (a, b) and (c, d) represent the same point if and only if we have $a = c$ and $b = d$. Points (a, b) with both a and b positive are said to lie in the first quadrant, those with $a < 0$ and $b > 0$ are in the second quadrant; and those with $a < 0$ and $b < 0$ are in the third quadrant; and those with $a > 0$ and $b < 0$ are in the fourth quadrant. Fig. 1-1 shows one point in each quadrant. The procedure for points in space is similar. We take three mutually perpendicular lines in space intersecting at a point (the origin). These lines determine three mutually perpendicular planes, and each point in space can be completely described by specifying, with appropriate regard for signs, its distances from these planes. For the present we confine our attention to plane analytic geometry.

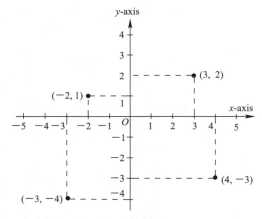

Fig. 1-1　Cartesian Coordinate System

1.2 Analytic Geometry

A geometric figure, such as a curve in the plane, is a collection of points satisfying one or more special conditions. By translating these conditions into expressions, involving the coordinates x and y, we obtain one or more equations which characterize the figure in question. For example, consider a circle of radius r with its center at the origin, as shown in Fig. 1-2. Let P be an arbitrary point on this circle, and suppose P has coordinates (x, y). Then the line segment OP is the hypotenuse of a right triangle whose legs have lengths $|x|$ and $|y|$ and hence, by the theorem of Pythagoras

$$x^2 + y^2 = r^2$$

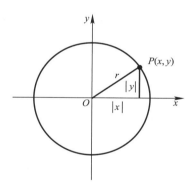

Fig. 1-2 Analytic Geometry

This equation, called a Cartesian equation of the circle, is satisfied by all points (x, y) on the circle and by no others, so the equation completely characterizes the circle. This example illustrates how analytic geometry is used to reduce geometrical statements about points to analytical statements about real numbers.

Throughout their historical development, calculus and analytic geometry have been intimately intertwined. New discoveries in one subject leads to improvements in the other. The development of calculus and analytic geometry is similar to the historical development, in that the two subjects are treated together. However, our primary purpose is to discuss calculus. Concepts from analytic geometry that are required for this purpose will be discussed as needed. Actually, only a few very elementary concepts of plane analytic geometry are required to understand the rudiments of calculus. A deeper study of analytic geometry is needed to extend the scope and applications of calculus, and this study will be carried out using the methods of calculus. Until then, all that is required from analytic geometry is a little familiarity with drawing graph of function.

Vocabulary and Expressions：

abscissa [æb'sisə] n. 横坐标

hypotenuse [hai'pɔtinju:z] n. 斜边，三角形的斜边，直角三角形的斜边

Theorem of Pythagoras [pai'θægərəs] 毕达哥拉斯定理，勾股定理

calculus ['kælkjuləs] n. 微积分，微积分学

analytic geometry 解析几何

horizontal [hori'zɔntl] adj. 水平的

calculation [ˌkælkju'leiʃn] n. 计算

Cartesian [kɑ:'ti:ziən] adj. 笛卡尔的；~ geometry 笛卡尔几何

Rene Descartes ['renei dei'ka:t] n. R. 笛卡尔（法国数学家、哲学家）

integral ['intigrəl] adj. 整数的，积分的；n. 积分

coordinate [kəu'ɔ:dinit] n. 坐标；~ system 坐标系；~ axis 坐标轴

intersect [intə'sekt] v. 相交

intertwine [intə'twain] v. 融合，结合

Cartesian coordinate system 笛卡尔坐标系

intimately ['intimitli] adv. 紧密地，亲密地

leg [leg] n. 侧边，直角边

ordered pair 有序对

ordinate ['ɔ:dinit] n. 纵坐标

origin ['ɔridʒin] n. 坐标原点

parabolic [ˌpærə'bɔlik] adj. 抛物线的

perpendicular [ˌpə:pən'dikjulə(r)] adj. （互

相）垂直的，正交的；n. 垂线

polygonal [pə'ligənl] adj. 多边形的

three-dimensional 三维的

triangle ['traiæŋgl] n. 三角形；right ~ 直角三角形

quadrant ['kwɔdrənt] n. 象限

reduce [ri'djuːs] v. 归结（为），化简

the unit distance 单位长度

vertical ['vəːtikəl] adj. 竖直的

Notes：

① R. 笛卡尔(1596~1650)提出了一种好得多的办法，并建立了解析（或笛卡尔）几何这个学科。

Exercises：

1. Translate the Following Paragraphs into Chinese.

(1) Let $A(x_1, y_1)$ be a fixed point on a straight line l with slope m. Let $P(x, y)$ represent any point on the line l, where $x \neq x_1$. Then the slope determined by P and A is equal to m. That is

$$\frac{y - y_1}{x - x_1} = m, \text{ or } y - y_1 = m(x - x_1)$$

The last equation is called the **_point-slope form_** of the equation of a line.

(2) A right angle is a 90° angle. An angle θ is acute if $0° < \theta < 90°$ or obtuse if $90° < \theta < 180°$. A straight angle is a 180° angle. Two acute angles are complementary if their sum is 90°. Two positive angles are supplementary if their sum is 180°.

2. Translate the Following Parapraphs into English.

(1) 数学培养学生分析问题的能力，使他们能应用毅力、创造性和逻辑思维推理来解决问题。

(2) 几何主要不是研究数，而是形，例如三角形、平行四边形和圆，虽然它也与数有关。

(3) 平面上的闭曲线中的每一点到一个固定点的距离均相等时叫做圆。这个固定点称为圆心，经过圆心且其两个端点在圆周上的线段称为这个圆的直径，直径的一半叫做半径，这条闭曲线的长度叫做周长。

(4) 对 xy 平面上的每一个点都指定了一个数对，称为它的坐标。

(5) 选取两条互相垂直的直线，其中一条是水平的，另一条是竖直的，把它们的交点记作 O，称为原点。

2　Function and Its Derivative
（函数及其导数）

教学目标：本单元的主要内容有函数概念、函数概念的历史发展由来、函数定义域和值域、映射关系、实值函数、恒等函数、绝对值函数、差商、极限、导数定义、零阶导数、一阶导数、二阶导数、n 阶导数、三角不等式以及导数的几何意义。教学重点是函数概念，难点是导数的几何意义——切线斜率。通过本单元的学习，要求理解数学家莱布尼茨的函数思想，掌握导数的定义和几何意义。

2.1　Function

Various fields of human have to do with relationships that exist between one collection of objects and another. Graphs, charts, curves, tables, formulas and Gallup polls are familiar to everyone who reads the newspapers. These are merely devices for describing special relations in a quantitative fashion. Mathematicians refer to certain types of these relations as functions. In this section, we give an informal description of the function concept.

E.g. 1. The force F necessary to stretch a steel spring a distance x beyond its natural length is proportional to x. That is, $F = cx$, where c is a number independent of x called the spring constant. This formula, discovered by Robert Hooke in the mid-17th century, is called Hooke's Law, and it is said to express the force as a function of the displacement.

E.g. 2. The volume of a cube is a function of its edge-length. If the edges have length x, the volume V is given by the formula $V = x^3$.

E.g. 3. A prime is any integer $n > 1$ that cannot be expressed in the form $n = ab$, where a and b are positive integers, both less than n. The first few primes are 2, 3, 5, 7, 9, 13, 17 and 19. For a given real number $x > 0$, it is possible to count the number of primes less than or equal to x. This number is said to be a function of x even though no simple algebraic formula is known for computing it (without counting) when x is known.

The word "function" is introduced into mathematics by Leibniz, who uses the term primarily to refer to certain kinds of mathematical formulas. *It is later realized that Leibniz's idea of function is much too limited in its scope, and the meaning of the word has since undergone many stages of generalization.* [1] Today, the meaning of function is essentially this: Given two sets, say X and Y, a function is a correspondence which associates with each element of X and only one element of Y. The set X is called the domain of the function. Those elements of Y associated with the elements in X form a set called the range of the function (This may be all of Y, but it need not be). Letters of the English and Greek alphabets are often used to denote functions. The particular letters f, g, h, F, G, H and φ are frequently used for this purpose. If f is a given function and if x is an object of its domain, the notation $f(x)$ is used to designate that object in the range which is associated to x by the function f; and it is called the value of f at x or the image of x under f. The symbol $f(x)$ is read as "f of x".

The function idea may be illustrated schematically in many ways. For example, in Fig. 2-1(a) the collections X and Y are thought of as sets of points and an arrow is used to suggest a "pairing" of a typical point x in X with the image point $f(x)$ in Y. Another scheme is shown in Fig. 2-1(b). Here the function f is imagined to be like a machine into which objects of the collection X are fed and objects of Y are produced. When an object x is fed into the machine, the output is the object $f(x)$. Although the function idea places no restriction on the nature of the objects in the domain X and in the range Y, in elementary calculus we are primarily interested in functions whose domain and range are sets of real numbers. Such functions are called real-valued functions of a real variable, or, more briefly, real functions, and they may be illustrated geometrically by a graph in the xy-plane. We plot the doman X on the x-axis, and above each point x in X we plot the point (x, y), where $y = f(x)$. The totality of such points (x, y) is called the graph of the function.

Now we consider some more examples of real functions.

E.g. 4. The identity function. Suppose that $f(x) = x$ for all real x. This function is often called the identity function. Its domain is the real line, that is, the set of all real numbers. Here $x = y$ for each point (x, y) on the graph of f. The graph is a straight line making equal angles with the coordinates axes

(Fig. 2-2a). The range of f is the set of all real numbers.

E.g. 5. The absolute-value function. Consider the function which assigns to each real number x the nonnegative number $|x|$. A portion of its graph is shown in Fig. 2-2(b). Denoting this function by φ, we have $\varphi(x) = |x|$ for all real x. For example, $\varphi(0) = 0$, $\varphi(2) = 2$, $\varphi(-3) = 3$. We list here some properties of absolute values expressed in function notation.

(1) $\varphi(-x) = \varphi(x)$
(2) $\varphi(x^2) = x^2$
(3) $\varphi(x+y) \leqslant \varphi(x) + \varphi(y)$ (the triangle inequality)
(4) $\varphi[\varphi(x)] = \varphi(x)$
(5) $\varphi(x) = \sqrt{x^2}$

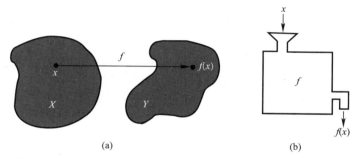

(a)　(b)

Fig. 2-1　Graphic Expression of Function Idea
(a) Mapping a Point x in X to the Image Point $f(x)$ in Y ; (b) An Input-Output Machine

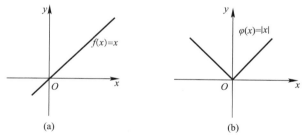

(a)　(b)

Fig. 2-2　Function Graph
(a) Identity Function $f(x)=x$; (b) Absolute-Value Function $\varphi(x)=|x|$

Seldom has a single concept played so important a role in mathematics as has the concept of function. It is desirable to know how the concept has developed. This concept, like many others, originates in physics. The physical quantities were the forerunners of mathematical variables, and relation among them was

called a function relation in the late 16th century. For instance, the formula $s = 16t^2$ for the number of meter s a body falls in any number of seconds t is a function relation between s and t, it describes the way s varies with t. The study of such relations led people in the 18th century to think of a function relation as nothing but a formula. Only after the rise of modern analysis in the early 19th century could the concept of function be extended. In the extended sense, a function may be defined as follows: If a variable y depends on another variable x in such a way that to each value of x corresponds a definite value of y, then y is a function of x. This definition serves many a practical purpose even today.

Not specified by this definition is the manner of setting up the correspondence. It may be done by a formula as the 18th century mathematics presumed, but it can equally well be done by tabulation such as a statistical chart, or by some other form of description. A typical example is the room temperature, which obviously is a function of time. But this function admits of no formula representation, although it can be recorded in a tabular form or traced out graphically by an automatic device. The modern definition of a function y of x is simply a mapping from a space X to another space Y. A mapping is defined when every point x of X has a definite image y, a point of Y. The mapping concept is close to intuition, and therefore desirable to serve as a basis of the function concept. Moreover, as the space concept is incorporated in this modern definition, its generality contributes much to the generality of the function concept.

2.2   The Derivative of a Function

The example described in the foregoing section points the way to the introduction of the concept of derivative. We begin with a function f defined at least on some open interval (a, b) on the x-axis. Then we choose a fixed point x in this interval and introduce the difference quotient

$$\frac{f(x + h) - f(x)}{h} \quad (2\text{-}1)$$

Where, the number h, which may be positive or negative (but not zero), is such that $x+h$ also lies in (a, b). The numerator of this quotient measures the change in the function when x changes from x to $x+h$. The quotient itself is referred to as the average rate of the change of f in the interval joining x to $x+h$. Now we let h approach zero and see what happens to this quotient. *If the quotient approaches some definite value as a limit (which implies that the limit is the same whether h approaches zero through positive values or through negative values), then this limit is called the derivative of f at x and is denoted by the symbol $f'(x)$ (read as "f prime of x").* [2] Thus, the formal definition of $f'(x)$ may be stated as follows:

The derivative $f'(x)$ is defined by the equation

$$f'(x) = \lim_{h \to 0} \frac{f(x + h) - f(x)}{h} \quad (2\text{-}2)$$

provided the limit exists. The number $f'(x)$ is also called the rate of change of f at x. We see that the concept of instantaneous velocity is merely an example of the concept of derivative. The velocity $v(t)$ is equal to the derivative $f'(t)$, where f is the function which measures position. This is often described by saying that velocity is the rate of change of position with respect to time.

In general, the limit process which produces $f'(x)$ from $f(x)$ gives us a way of obtaining a new function f' from a given function f. [3] The process is called differentiation, and f' is called the first derivative of f. If f', in turn, is defined on an open interval, we can try to compute its first derivative, denoted by f'' and called the second derivative of f. Similarly, the nth derivative of f, denoted by $f^{(n)}$, is defined to be the first derivative of $f^{(n-1)}$. We make the convention that $f^{(0)}=f$, that is, the 0th derivative is the function itself.

The procedure used to define the derivative has a geometric interpretation which leads in a natural way to the idea of a tangent line to a curve. A portion of the graph of a function f is shown in Fig. 2-3. Two of its points P and Q are shown with respective coordinates $(x, f(x))$ and $(x+h, f(x+h))$. Consider the right triangle with hypotenuse PQ; its altitude, $f(x+h) - f(x)$, represents the difference of the ordinates of the two points Q and P. Therefore, the difference quotient Eq. (2-1) represents the trigonometric tangent of the angle α that PQ makes with the horizontal. The real number $\tan\alpha$ is called the slope of the line through P and Q and it provides a way of measuring the "steepness" of this line. For example, if f is a linear function, say $f(x) = mx + b$, the difference quotient Eq. (2-1) has the value m, so m is the slope of the line.

that f has a derivative at x. This means that the difference quotient approaches a certain limit $f'(x)$ as h approaches 0. *When this is interpreted geometrically, it tells us that, as h gets nearer to 0, the point P remains fixed, Q moves along the curve toward P, and the line through PQ changes its direction in such a way that its slope approaches the number $f'(x)$ as a limit.* [④] For this reason it seems natural to define the slope of the curve at P to be the number $f'(x)$. The line through P having this slope is called the tangent line at P.

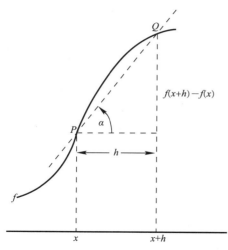

Fig. 2-3 Geometric Interpretation of the Difference Quotient as the Tangent of an Angle

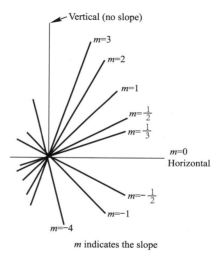

m indicates the slope

Fig. 2-4 Lines of Various Slopes

Some examples of lines of various slopes are shown in Fig. 2-4. For a horizontal line, $\alpha = 0$ and the slope, $\tan\alpha$, is also 0. If α lies between 0 and $\pi/2$, the line is rising as we move from left to right and the slope is positive. If α lies between $\pi/2$ and π, the line is falling as we move from left to right and the slope is negative. A line for which $\alpha = \pi/4$ has slope 1. As α increases from 0 to $\pi/2$, $\tan\alpha$ increases without bound, and the corresponding lines of slope $\tan\alpha$ approach a vertical position. Since $\tan(\pi/2)$ is not defined, we say that vertical lines have no slope. Suppose now

Vocabulary and Expressions：

in a quantitative fashion 在数量上

tabulation ［ˌtæbjuˈleiʃən］ n. 表格，列表，制表

the absolute-value function 绝对值函数

generalization ［ˌdʒenərəlaiˈzeiʃən］ n. 推广，一般化

Hooke's Law 胡克定律

Robert Hooke ［ˈrɔbət huk］ n. R. 胡克（物理学家）

cube ［kjuːb］ n. 立方体

the identity function 恒等函数

displacement ［disˈpleismənt］ n. 位移

image ［ˈimidʒ］ v. 想象；n. （映射的）像，图像

domain ［dəuˈmein］ n. 区域，定义域

edge [edʒ] n. 棱，边

Leibniz ['libniz] n. 莱布尼茨(数学家)

function idea 函数思想

limit ['limit] v. 限制；n. 极限

Gallup poll ['gæləp pəul] 盖洛普民意测验

nonnegative [nɔn'negətiv] adj. 非负的

plot [plɔt] v. 画(草图)

schematic representation 图解表示

prime [praim] n. 素数，质数

schematically [ski'mætikəli] adv. 图解式地

proportional [prə'pɔ:ʃənəl] adj. 成比例的

range [reindʒ] n. 值域，范围

spring constant 劲度系数(或弹簧弹性系数)

stretch [stretʃ] v. 拉伸

the real-valued function 实值函数

totality [təu'tæliti] n. 全部，全体

the triangle inequality 三角不等式(即在任何三角形中，任意两边之和大于第三边。或 $|a|-|b| \le |a+b| \le |a|+|b|$)

admit [ad'mit] v. 准许；~ of 容许；~ of no 不容许

intuition ['intju'iʃən] n. 直观

inverse function 反函数

mapping ['mæpiŋ] n. 映射

characteristic function 特征函数

measurable ['meʒərəbl] adj. 可测的

extended [iks'tendid] adj. 广义的

finitely many 有限多个

parabola [pə'ræbələ] n. 抛物线

forerunner ['fɔ:,rʌnə] n. 先行者

presume [pri'zju:m] v. 假定

simple function 简单函数

in term of y 用 y 来表示

derivative [di'rivətiv] n. 导数；the first ~ 一阶导数；the second ~ 二阶导数

differentiation [,difə,renʃi'eiʃən] n. 微分法

approach zero 趋于零

open interval 开区间

linear function 线性函数

without bound 无界，无限

numerator ['nju:məreitə] n. 分子

constant [kɔnstənt] n. 常数

rectilinear motion 直线运动

instantaneous velocity 瞬时速度

slope [sləup] n. 斜率，坡度

Notes：

① 这里 it 是形式主语，代替 that 引起的主语从句。本句可译成：后来人们才认识到，莱布尼茨的函数思想适用的范围太过局限了，这个术语的含义从那时起经过了多次推广。

② 如果差商以某个确定的值为极限(这蕴涵着不论 h 取正的值趋于 0 还是取负的值趋于 0，其极限一样)，那么这个极限称为 f 在 x 的导数，记作 $f'(x)$(读成"f一撇 x")。

③ 一般地，由 $f(x)$ 产生 $f'(x)$ 的极限过程向我们提供了一种从一个给定的函数 f 得到一个新的函数 f' 的方法。

④ 其几何意义为，当 h 趋于 0 时，点 P 保持不动，而点 Q 沿曲线趋近 P；同时，经过 PQ 的直线不断地改变方向，结果其斜率趋于数值 $f'(x)$，并以它为极限。

Exercises：

1. Translate the Following Paragraphs into Chinese.

(1) We can use the concepts of ordered pairs to give a new definition of functions as follows. A function is a set of ordered pairs such that for each first coordinate only one second coordinate exists. The domain of a function is the set of all first coordinates. The range of a function is the set of all second coordinates.

(2) Let X and Y be sets and suppose $f : X \to Y$ is a function. If $g : Y \to X$ is another function and has the property that

$$y = f(x) \text{ if and only if } x = g(y)$$

then we call g the inverse function to f. Observe that $x = g(y)$ is what we obtained by solving the equation $y = f(x)$ for x in term of y. However, in the general case, the equation $y = f(x)$ may have no solutions at all or else may have many solutions. Thus, for $f : X \to Y$ to admit an inverse function, it is necessary that, for each y in the set of Y, the equation $y = f(x)$ has a unique solution x in the set X.

(3) Examples of derivatives：

E.g. 1：Derivative of a constant function. Suppose f is a constant function, say $f(x) = c$ for all x. The difference quotient is

$$\frac{f(x + h) - f(x)}{h} = \frac{c - c}{h} = 0$$

Since the quotient is 0 for all $h \neq 0$, its limit, $f'(x)$ is also 0 for every x. In the other words, a constant function has a zero derivative everywhere.

E.g. 2：Derivative of a linear function. Suppose f is a linear function, say $f(x) = mx + b$ for all real x. If $h \neq 0$, we have

$$\frac{f(x + h) - f(x)}{h}$$
$$= \frac{m(x + h) + b - (mx + b)}{h}$$
$$= \frac{mh}{h}$$
$$= m$$

Since the difference quotient does not change when h approaches 0, we conclude that

$$f'(x) = m \qquad \text{for every } x$$

Thus, the derivative of a linear function is a constant function.

2. Translate the Following Paragraphs into English.

（1）若 f 是一个给定的函数，x 是定义域里的一个元素，那么记号 $f(x)$ 用来表示由 f 确定的对应于 x 的值。

（2）速度等于位置函数的导数。

（3）在直线运动中，速度的一阶导数称为加速度。

（4）当 α 从 0 增加到 $\pi/2$ 时，$\tan\alpha$ 无限增加，而 $\tan\alpha$ 所对应的直线趋于竖直位置。

3　Computer Hardware
（计算机硬件）

教学目标：本单元的主要内容有计算机硬件，中央处理器 CPU 芯片，内存，只读存储器 ROM，随机存储器 RAM，输入输出设备如键盘、鼠标、扫描仪、麦克风、显示器（分辨率、点距、刷新率）、打印机、音箱、调制解调器及系统总线。教学重点是输入输出设备，难点是 CPU 芯片和调制解调器。通过本单元的学习，要求熟悉计算机硬件主要由哪些部件组成，每个部件有什么功能。特别要理解调制解调器可以将数据从数字信号转换成可以通过电话线传输的模拟信号。

A computer is a fast and accurate symbol manipulating system that is organized to accept, store and process data and produce output results under the direction of a stored program of instructions. Fig. 3-1 shows the basic organization of a computer system. Key elements in this system include CPU (Central Processing Unit), the memory, I/O (input and output) devices. Let's examine each component of the system in more detail.

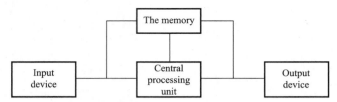

Fig. 3-1   The Basic Organization of a Computer System

3.1   CPU

A processor is a functional unit that interprets and carries out instructions. Every processor comes with a unique set of operations such as ADD, STORE or LOAD that represent the processor's instruction set. Computer designers are fond of calling their computer machines, so the instruction set is sometimes referred to as machine instructions and the binary language in which they are written is called machine language. You should not confuse the processor's instruction set with the instructions found in high-level programming languages, such as BASIC or PASCAL. The processor is the "brains" of the computer that has the ability to carry out our instructions or programs given to the computer. The processor is the part that knows how to add and subtract and to carry out simple logical operations. In a big mainframe computer the processor is called a Central Processing Unit, or CPU, while in a microcomputer, it is usually known as a microprocessor. There are two main sections found in the CPU of a typical personal computer system: the arithmetic-logic section and the control section. But these two sections are not unique to personal computer: They are found in CPUs of all sizes.

An instruction is made up of operations that specify the function to be performed and operands that represent the data to be operated on. For example, if an instruction is to perform the operation of adding two numbers, it must know: (1) What the two numbers are. (2) Where the two numbers are. When the numbers are stored in the computer's memory, they have an address to indicate where they are, so if an operand refers to data in the computer's memory it is called an address. The processor's job is to retrieve instructions and operands from memory and to perform each operation. Having done that, it signals memory to send it the next instruction. This step-by-step operation is repeated over and over again at awesome speed. The CPU means the Central Processing Unit. It is the heart of a computer system (Fig. 3-2). The CPU in a microcomputer is actually one relatively small integrated circuit or chip. *Although most CPU chips are smaller than a lens of a pair of glasses, the electronic components they contain would have filled a room a few decades ago.* [①] Using advanced microelectronic techniques, manufacturers can cram tens of thousands of circuits into tiny layered silicon chips that work dependably and use less power.

Fig. 3-2   CPU

The CPU coordinates all the activities of the various components of the computer. It determines which operations should be carried out and in what order. The CPU can also retrieve information from memory and can store the results of manipulations back into the memory unit for later reference. The basic job of computers is the processing of information. For this reason, computers can be defined as devices which accept information in the form of instructions called a program and characters called data, perform mathematical and logical operations on the information, and then supply results of these operations. *The program, which tells the computers what to do, and the data, which provide the information needed to solve the problem, are kept inside the computer in a place called memory.* [②]

Computers are thought to have many remarkable powers. However, most computers, whether large or small, have three basic capabilities. First, computers have circuits for performing arithmetic operations, such as: addition, subtraction, multiplication, division and exponentiation. Second, computers have a means of communicating with the user. *After all, if we couldn't feed information in and get results back, these machines would not be of much use.* [③] Third, computers have circuits which can make decisions. The kinds of decisions which computer circuits can make are of the type: Is one number less than another? Are two numbers equal? And, is one number greater than another? A processor is composed of two functional units: a control unit and an arithmetic/logic unit, and a set of special workspaces called registers.

3.2   The Memory

The memory (Fig. 3-3) is the computer's work area. There are two types of memory chips: Read Only Memory (ROM) and Random Access Memory (RAM). ROM chips are designed for applications in which data is only read (This data can include program instructions). These chips are programmed with data by an external programming unit before they are added to the computer system. Once this is done, the data usually does not change. A ROM chip alway sretains its data, even when power to the chip is turned off. RAM also called read/write memory, can be used to store data that changes. Unlike ROM, RAM chips lose their data once power is shut off. Many computer systems, including personal computers, include both ROM and RAM.

Fig. 3-3   The Memory

3.3   I/O Devices

A computer is a powerful machine that can do everything people assign it. However the computer can't communicate with people directly. By the aid of input and output devices, a computer and people can "know" each other. Using input devices, people "tell" the computer what it should do and the computer feeds back the result through output devices. Most of the input devices work in similar ways. The messages or signals received are encoded into patterns which CPU can process by input devices, then, conveyed to CPU. Input devices can not only deliver information to CPU but also activate or deactivate processing just as light switches turn lamps on or off. Output devices can tell the processing results and warn users where their programs or operations are wrong. The most common output devices are monitor, matrix printer, inkjet printer, laser printer, plotter for drawing, speaker, etc. They also work in similar ways. They decode the coded symbols produced by CPU into forms of information that users understand or use easily and show them. Input and output devices are the interfaces of man and machine. They usually include keyboard, mouse, monitor, printer, disk (hard disk or floppy disk), input pen, scanner, microphone, etc. Let's consider some input and output devices that are in common use.

(1) Keyboard: The keyboard is used to type information into the computer or input information (Fig. 3-4). There are many different keyboard layouts and sizes with the most common for Latin based languages being the QWERTY layout (named for the first six keys). The standard keyboard has 101 keys. Notebooks have embedded keys accessible by special keys or by pressing key combinations. Some of the keys on a standard keyboard have a special use. There are referred to as command keys. The three most common are the Control or Ctrl, Alternate or Alt and the Shift keys. Each key on a standard keyboard has one or two characters. Press the key to get the lower character and hold Shift to get the upper. The numeric keypad is located on the right side of the keyboard and looks like an adding machine. However, when you are using it as a calculator, be sure to depress the Num Lock key so the light above Num Lock is lit.

Fig. 3-4   Keyboard

The function keys (F1, F2 and so forth) are usually located at the top of the keyboard. These keys are used to give the computer commands. The function of each key varies with each software program. *The arrow keys allow you to move the position of the cursor on the screen.* [4] Special-purpose keys perform a specialized function. The Esc key's function depends on the program being used. Usually it will back you out of a command. The Print Screen sends a copy of whatever is on the screen to the printer. The Scroll Lock key, which does not operate in all programs, is rarely used with today's software. The Num Lock key controls the use of the number keypad. The Caps Lock key controls typing text in all capital letters.

(2) Mouse: A mouse is a small device that a computer user pushes across a desk surface in order to point to a place on a display screen and to select one or more actions to take from that position (Fig. 3-5). The most conventional kind of mouse has two buttons on the top: the left one is used most frequently. In the Windows operating systems, it allows the user to click once to send a "Select" indication that provides the user with feedback that a particular position

has been selected for further action. The next click on a selected position or quick clicks on it causes a particular action to take place on the selected object. The second button, on the right, usually provides some less-frequently needed capability. For example, when viewing a web page, you can click on an image to get a pop-up menu where, among other things, you can save the image on your hard disk.

Fig. 3-5　Mouse

A mouse consists of a metal or plastic housing or casing, a ball that sticks out of the bottom of the casing and rolls on a flat surface, one or more buttons on the top of the casing, and a cable that connects the mouse to the computer. [5] As the ball is moved over the surface in any direction, a sensor sends impulses to the computer that causes a mouse-responsive program to reposition a visible indicator (called a cursor) on the display screen. The positioning is relative to some starting place. Viewing the cursor's present position, the user readjusts the position by moving the mouse.

(3) Scanner: With the progress of the technique of input and output devices, scanners (Fig. 3-6) gradually come into the ordinary family these years. They can input any kinds of information which are printed on a paper with the most convenient way, by the aid of recognizing and analyzing software, all will be scanned into the computer within several minutes.

(4) Microphone: Have you ever tried to control computer through your voice? Voice control, with the aid of microphone, is already used in some very popular word-processing software.

(5) Monitor: The monitor shows information on the screen when you type. This is called

outputting information. When the computer needs more information, it will display a message on the screen, usually through a dialog box. Monitors come in many types and sizes from the simple monochrome (one color) screen to full color screens (Fig. 3-7). A character-based display divided the screen into a grid of rectangles, each of which can display a single character. The set of characters that the screen can display is not modifiable; therefore, it is not possible to display different sizes or styles of characters. A bitmap display divides the screen into a matrix of tiny, square "dots" called pixels. Any characters or graphics that the computer displays on the screen must be constructed of dot patterns within the screen matrix. The more dots your screen displays in the matrix, the higher its resolution.

Fig. 3-6　Scanner

Fig. 3-7　Monitor

(a) Resolution: Resolution refers to the number of individual dots of color, known as pixels, contained on a display. Resolution is typically expressed by identifying the number of pixels on the horizontal axis (rows) and the number on the vertical axis (columns), such as

640×480. The monitor's viewable area, refresh rate and dot pitch all directly affect the maximum resolution a monitor can display.

(b) Dot pitch: Briefly, the dot pitch is the measure of how much space there is between a display's pixels. When considering dot pitch, remember that the smaller the better. Packing the pixels closer together is fundamental to achieving higher resolutions. A display normally can support resolutions that match the physical dot (pixel) size as well as several lesser resolutions.

(c) Refresh rate: In monitors based on CRT technology, the refresh rate is the number of times that the image on the display is drawn each second. Refresh rates are very important because they control flicker, and you want the refresh rate as high as possible.

(6) Printer: A common type of printer (Fig. 3-8) is the matrix printer. A cheap printer might have seven needles, for printing 80 characters in 5×7 matrix across the line. In effect, the print line then consists of 7 horizontal lines, each consisting of 5×80 = 400 dots. Each dot can be printed or not printed, depending on the characters to be printed. The print quality can be increased by two techniques: using more needles and having the circles overlap. The ink jet printer uses a nozzle and sprays ink onto the paper to form the appropriate characters. In order to get the correct character, the ink is directed with a valve and one or more electronic deflectors that control the vertical and horizontal position of the jet stream of ink. It is possible to print a number of different characters with different styles. Some ink jet printers are capable of printing images in full color. The laser printer uses laser beams that strike laser-sensitive paper. This paper then picks up a powder or a toner and the powder or toner is bonded to the paper by heat, pressure or both. With a laser printer, it is possible to print an entire page at one time. One of the disadvantages of many laser printers is the problem of static electricity. With some of these printers, the paper can have a tendency to stick together. This makes it difficult to output from a laser printer in developing bills and statements sent to customers. A newer cold laser printer has been developed to avoid this problem. They don't require heat in bonding the characters to the paper. As a result, the problems of static electricity and having the paper cling together can be eliminated or substantially reduced.

Fig. 3-8　Printer

(7) Sound box: *If software has music with it, you need a sound box which is attached to your computer to play the music.*[⑥] You may work on the computer while listening to the music. That is really an enjoyment.

(8) Modem: The need to communicate between distant computers leads to the use of the existing phone network for data transmission. Most phone lines are designed to transmit analog information—voices, while the computers and their devices work in digital form—pulses. *So, in order to use an analog medium, a converter between the two systems is needed.*[⑦] This converter is the modem (Fig. 3-9).

Fig. 3-9　Modem

A modem is a device that converts data from

digital computer signals to analog signals that can be sent over a phone line. ⑧ This is called modulation. The analog signals are then converted back into digital data by the receiving modem. This is called demodulation. Modems can be classified into external ones and internal ones. Typically, external modems feature an array of lights set in a display panel that offers important information when you are trying to troubleshoot your setup. You also need a correctly wired cable to connect your modem to an available serial port on your computer. Internal modems are printed circuit boards that take up one of the available expansion slots inside of your computer.

3.4 System Buses

The components of the computer are connected to the buses. To send information from one component to another, the source component outputs data onto the bus. The destination component then inputs this data from the bus. As the complexity of a computer system increases, it becomes more efficient (in terms of minimizing connections) at using buses rather than direct connections between every pair of devices. Buses use less space on a circuit board and require less power than a large number of direct connections. They also require fewer pins on the chip or chips that comprise the CPU.

Data is transferred via the data bus. When the CPU fetches data from memory, it first outputs the memory address on its address bus. Then memory outputs the data onto the data bus; the CPU can then read the data from the data bus. When writing data to memory, the CPU first outputs the address onto the address bus, and then outputs the data onto the data bus. The memory then reads and stores the data at the proper location. The processes for reading data from and writing data to the I/O devices are similar. The control bus is different from the other two buses. The address bus consists of n lines, which combine to transmit one n-bit address value. Similarly, the lines of the data bus work together to transmit a single multi-bit value. In contrast, the control bus is a collection of individual control signals. These signals indicate whether data is to be read into or written out of the CPU, whether the CPU is accessing memory or an I/O device, and whether the I/O device or memory is ready to transfer data. The control bus is really a collection of (mostly) unidirectional signals. Most of these signals are output from the CPU to the memory and I/O subsystems, although a few are output by these subsystems to the CPU.

Vocabulary and Expressions：

Central Processing Unit （CPU）中央处理器，中央处理单元

chip ［chip］ n. 芯片

memory ［'meməri］ n. 存储器，内存

memory bar 内存条

Input/Output（I/O）Device 输入输出设备

ENIAC-vacuum tube 埃尼亚克电子管（或真空管）

exponentiation ［,ekspə,nenʃi'eiʃən］ n. 幂运算

megahertz ［'megəhə:ts］ n. 兆赫（兹）

acronym ［'ækrənim］ n. 首字母缩略词

gateway ［'geitwei］ n. 网关

kilobyte （KB）［'kiləbait］ n. 千字节

megabyte （GB）［'megəbait］ n. 兆字节

modulation ［,mɔdju'leiʃn］ n. 调制

demodulation ［'di:,mɔdju'leiʃn］ n. 解调

feedback ［'fi:dbæk］ n. 反馈，回馈

feed back 反馈，回馈

ink jet(orink-jet, orinkjet) printer 喷墨打印机

microphone ［'maikrəfəun］ n. 麦克风，话筒

modem ［'məudem］（modulator-demodu lator）n. 调制解调器

monitor ［'mɔnitə(r)］ n. 显示器

resolution ［,rezə'lu:ʃn］ n. 分辨率

dot pitch 点距
refresh rate 刷新率
matrix printer 点阵打印机
keyboard ['ki:bɔ:d] n. 键盘；standard ~ 标
　准键盘
mouse [maus] n. 鼠标
Read Only Memory (ROM)只读存储器
Random Access Memory (RAM)随机存储器

Notes：
① 虽然大多数 CPU 芯片比一块眼镜片还小，但所包含的电子元件在几十年前却要装满一个房间。
② 程序的作用是指示计算机如何工作，而数据则是为解决问题提供的所需要的信息，两者都存储在存储器里。
③ 如果我们不能输入信息和取出结果，这种计算机毕竟不会有多大用处。
④ 方向键允许你移动光标在屏幕上的位置。
⑤ 鼠标由以下几个部分组成：一个金属或塑料的盒体，一个凸出于盒体底部并可以在平面上滚动的球体，位于盒体上部的一个或多个按键以及一条连接到计算机的电缆线。
⑥ 如果软件带有音乐，你就需要一个音箱连接到计算机上来播放。
⑦ 因此，为了利用传输模拟信号的媒介，在两种系统之间需要一个转换器。
⑧ 调制解调器可以将数据从数字信号转换成可以通过电话线传输的模拟信号。

Exercises：
1. Translate the Following Phrases into English or Chinese.
(1) machine instructions
(2) RAM bars
(3) expanded memory
(4) standard keyboard
(5) function keys
(6) capital letters
(7) pop-up menu
(8) 系统时钟
(9) 微电子技术
(10) 软件中断
(11) 存储器芯片
(12) 静电
(13) 冷激光打印机
(14) 外置式调制解调器

2. Translate the Following Paragraphs into Chinese.
(1) Computers are thought to have many remarkable powers. However, most computers, whether large or small, have three basic capabilities. First, computers have circuits for performing arithmetic operations, such as: addition, subtraction, multiplication, division and exponentiation. Second, computers have a means of communicating with the user. After all, if we couldn't feed information in and get results back, these machines would not be of much use. Third, computers have circuits which can make decisions. The kinds of decisions which computer circuits can make are of the type: Is one number less than another? Are two numbers equal? And, is one number greater than another? A processor is composed of two functional units: a control unit and an arithmetic/logic unit, and a set of special workspaces called registers.
(2) The memory is the computer's work area. There are two types of memory chips: Read Only Memory (ROM) and Random Access Memory (RAM). ROM chips are designed for applications in which data is only read (This data can include program instructions). These chips are programmed with data by an external programming unit before they are added to the computer system. Once this is done, the data usually does not change. A ROM chip always retains its data, even when power to the chip is turned off. RAM also called read/write memory, can be used to store data that changes. Unlike ROM, RAM chips lose their data once power is shut off. Many computer systems, including personal computers, include both ROM and RAM.
(3) Resolution: Resolution refers to the number of individual dots of color, known as pixels, contained on a display. Resolution is typically expressed by identifying the number of pixels on the horizontal axis (rows) and the number on the vertical axis (columns), such as 640×480. The monitor's viewable area, refresh rate and dot pitch all directly affect the maximum resolu-

tion a monitor can display.

（4）Dot Pitch：Briefly, the dot pitch is the measure of how much space there is between a display's pixels. When considering dot pitch, remember that the smaller the better. Packing the pixels closer together is fundamental to achieving higher resolutions. A display normally can support resolutions that match the physical dot (pixel) size as well as several lesser resolutions.

（5）Refresh Rate：In monitors based on CRT technology, the refresh rate is the number of times that the image on the display is drawn each second. Refresh rates are very important because they control flicker, and you want the refresh rate as high as possible.

3. Translate the Following Paragraphs into English.

（1）计算机由许多被称为硬件的部分组成。硬件就是计算机中能看见和触摸的部分。

（2）显示器是最普通的一种输出设备，用来将计算机正在做的事显示给用户。

（3）计算机的键盘非常敏感，你只需轻轻触摸一下就可以打出字来。

4 OA
（办公自动化）

教学目标：本单元的主要内容有办公自动化技术、文字处理系统、电子表格 Excel、数据库管理系统、图形图像处理、视频会议（远程会议）、E-mail、语音识别与集成以及集成信息处理网络。教学重点是图形图像处理和视频会议，难点是电子表格和语音识别与集成。通过本单元的学习，要求熟悉常用的计算机办公自动化技术，掌握收发 E-mail、文字处理系统、电子表格 Excel、图形图像处理（如 CAD、Photoshop）以及视频会议（如腾讯会议、ZOOM 会议）等。

OA（Office Automation）is the application of computer and communications technology to improve the productivity of clerical and managerial office workers. In the mid-1950s, the term was used as a synonym for almost any form of data processing, referring to the ways in which bookkeeping tasks were automated. *After some years of disuse, the term was revived in the mid-1970s to describe the interactive use of word and text processing systems, which would later be combined with powerful computer tools, thereby leading to a so-called "integrated electronic office of the future".*[1] Personal computer-based office automation software has become an indispensable part of electronic management in many countries. Word processing programs have replaced typewriters; spreadsheet programs have replaced ledger books; database programs have replaced paper-based electoral rolls, inventories and staff lists; personal organizer programs have replaced paper diaries; and so on.

Office automation encompasses six major technologies: (1) Data processing—Information in numeric form usually calculated by a computer; (2) Word processing—Information in text form—words and numbers; (3) Graphics—Information that may be in the form of numbers and words, then keyed into a computer and displayed on a screen in a graph, chart, table or other visual form that makes it easier to understand; (4) Image—Information in the form of pictures. Here an actual picture or photograph is taken, entered into the computer, and shown on a screen; (5) Voice—The processing of information in the form of spoken words; (6) Networking—The linking together electronically of computer and other office equipment for processing data, words, graphics, image and voice.

Initially, systems sold by major manufacturers are aimed at clerical and secretarial personnel. These are mainly developed to do word processing and record processing (maintenance of small sequential files, such as names and addresses, which are ultimately sorted and merged into letters). More recently, attention has also been focused on systems (Fig. 4-1), which directly support the principals (managers and professional workers). Such systems emphasize the managerial communications function. Today's organizations have a wide variety of office automation hardware and software components at their disposal. The list includes telephone and computer systems, electronic mail, word processing, desktop publishing, database management systems, two-way cable TV, office-to-office satellite broadcasting, on-line database services and voice recognition and synthesis. Each of these components is intended to automate a task or function that is presently performed manually. But experts agree that the key to attaining office automation lies in integration—incorporating all the components into a whole system such that information can be processed and communicated with maximum technical assistance and minimum human intervention.

The fast pace of modern business requires critical information quickly. At the same time, government demands and business bureaucracy require extensive amounts of paperwork. *As a result, modern business offices are reexamining traditional methods of doing office work to find better ways to capture and communicate information when and where it is needed.*[2] They seek the most efficient method to generate, record, process, file and communicate or distribute information. Modern technology offers office automation as the foundation of an economical solution. Present and future office system aim to develop integrated information processing networks that bring together everything a firm needs to conduct its daily business effectively.

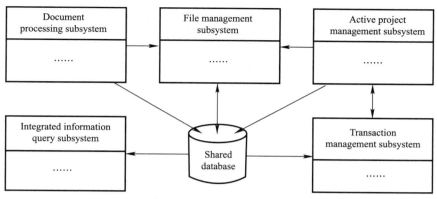

Fig. 4 -1　Office Automation System

4. 1　Word Processing

Word processing refers to the methods and procedures involved in using a computer to create, edit and print documents. Most standard word processing features are supported, including footnotes and mail-merge but no tables or columns. The interface uses customizable toolbars, and the editing screen is a zoom-able draft mode that optionally displays headers, footnotes and footers. An un-editable print preview displays a full-page or facing-page view. *Fonts, keyboard layouts and input direction change when you select a new language, and keyboard layouts can be customized.* ③

Word processors vary considerably, but all word processors support some basic features. Word processors that support only the basic features (and maybe a few others) are called text editors. Most word processors, however, support additional features that enable you to manipulate and format documents in more sophisticated ways. These more advanced word processors are sometimes called full-featured word processors. Full-featured word processors usually support the following features: (1) Insert and delete text; (2) Cut, paste and copy; (3) Page size and margins; (4) Search and replace; (5) Word wrap; (6) Print; (7) File management; (8) Font specifications; (9) Footnotes and cross-references; (10) Graphics; (11) Headers, footers and page numbering; (12) Layout; (13) Macros; (14) Merge; (15) Spell checker; (16) Tables of contents and indexes; (17) Thesaurus; (18) Windows; (19) WYSIWYG (What you see is what you get).

4. 2　Electronic Spreadsheet

Microsoft Excel is a spreadsheet program that allows you to organize data, complete calculations, make decisions, graph data, develop professional looking reports, publish organized data on the web and access real-time data from web sites. ④ Using Microsoft Excel, you can create hyperlinks within a worksheet to access other Office documents on the network, an organization's intranet or the Internet. Electronic spreadsheet can be used in all walks of life, for example, accountants use electronic spreadsheets to check financial statements and compile payrolls; in commerce, electronic spreadsheets are used to prepare budget and perform comparison on quoted prices; teachers use electronic spreadsheets to record marks of courses students get in examinations; scientists use electronic spreadsheets to analyze experimental data; housewives use electronic spreadsheets to keep track of family expenditure.

In Excel, formulas and functions can be used to perform statistics or computations of data in electronic spreadsheets. When data change, the results of computations will automatically upgrade. All formulas begin with the sign "=". It contains arithmetic operation signs, text operation signs, comparative operation signs and quoting operation signs—totally four categories of signs. Functions are built-in formulas pre-defined in Excel, including SUM, AVERAGE, COUNT, MAX, etc. Various statistic charts in electronic spreadsheets represent the data in unit squares, so as to make the data easy and intuitive to understand, meanwhile, when the data change, the charts automatically follow the change.

In Excel, charts include "imbedded chart" and "independent chart". The former is placed and displayed and printed together with worksheets; the latter is an independently generated worksheet, and it is printed separated from the original datasheet. In Excel, there are two-dimensional chart and three-dimentational charts. They are gradually generated by using the "chart direction" button in toolbar or "picture" command in "insert" menu. The user can edit the generated charts. There are zigzag chart, bar chart, pie chart and others available for selection. Excel has not only abilities to perform simple data management and computation but also to set up database. To use datasheet, increase or delete record, and do sequencing, sieving, classifying and summing up data.

4.3　Video Conferencing

Video conferencing involves the linking of remote sites by one-way or two-way television. *If conference rooms in offices can be equipped with the necessary audiovisual facilities, travel time and money can be saved by holding a teleconference instead of a face-to-face conference.*[5] Many businesses are experimenting with sales and board meetings through video conferencing. The cost is still high, especially if the video conference involves a direct two-way satellite channel connection.

Vocabulary and Expressions:

clerical ['klerikl] adj. 办公室工作的，办事员的，书记的，牧师的

bookkeeping ['bukki:piŋ] n. 簿记，记账，统计

ledger book 分类账簿

electoral [i'lektərəl] adj. 选举的，选举人的

thesaurus [θi'sɔ:rəs] n. 词库，辞典，宝库，

知识宝库，分类词汇汇编，百科全书

video conferencing 视频会议，电视会议

footnote ['futnəut] n. 脚注

footer ['futə(r)] n. 页脚

zoom-able, zoomable ['zu:məbl] adj. 可缩放的，可调整大小的

audiovisual [ˌɔ:diəu'viʒuəl] adj. 视听的，视听教学的；n. 视听教材

teleconference ['telikɔnfrəns] n. [通信]远程会议，电话会议

Notes:

① 经若干年搁置后，20 世纪 70 年代中期该词再次被用来描述字和文本处理系统的交互使用，这种系统后来又与强有力的计算机结合导致所谓的"未来综合电子办公室"的出现。

② 因此，现代商业办公室都在对传统的办公方法进行研究与调整，以找到可以随时随地获取信息和传递信息的较好的途径。

③ 当你选择一种新语言时，字体、键盘布局和输入方向均可变化，并且键盘布局可以定制。

④ Excel 是一个电子表格程序，用于组织数据、完成计算、做出决策、将数据图表化、生成专业水准的报告、在 Web 上发布组织好的数据以及在 Web 站点上存取实时数据。

⑤ 如果会议室内能安装所需的声像设备，通过举行远程会议可代替面对面的会议，可节约旅行所花费的时间和经费。

Exercises：

1. Translate the Following Paragraphs into Chinese.

（1）Today's organizations have a wide variety of office automation hardware and software components at their disposal. The list includes telephone and computer systems, electronic mail, word processing, desktop publishing, database management systems two-way cable TV, office-to-office satellite broadcasting, on-line database services and voice recognition and synthesis. Each of these components is intended to automate a task or function that is presently performed manually. But experts agree that the key to attaining office automation lies in integration—incorporating all the components into a whole system such that information can be processed and communicated with maximum technical assistance and minimum human intervention.

（2）The arrival of the digital technologies has been a boon to the educational field, and has led, in recent years, to the institutions of higher learning in China rapidly embracing digital multimedia technology in their educational curricula. The use of modern technology on beefing up the delivery of learning materials in our education system must reflect the changing times. Consequently the educators at higher institutions of learning are facing a new challenge today, i.e., to integrate these multimedia technologies into the classroom to enhance the teaching and learning environments for both the teacher and the students.

2. Translate the Following Paragraphs into English.

（1）如果你有不止一个文件打开着并且均在屏幕上可见，你就能使用拖放编辑将文件从一个文档移到另一个文档。

（2）拖放编辑适合在短距离内重新放置文本，但使用剪切、复制和粘贴，你可以在文档之间长距离移动文本或图像，或是生成文本的拷贝而不是移动原始拷贝。

（3）要粘贴文本或图像，首先将光标插入点放置到你需要粘贴项出现的位置，然后从编辑菜单中选择粘贴命令，或按 Ctrl + V，或单击标准工具栏中的粘贴按钮。

5　Stress, Strain and Hooke' s Law
（应力、应变与胡克定律）

教学目标： 本单元的主要内容有横截面、隔离体、拉力、压力、应力定义、拉应力、压应力、应变定义、拉应变、压应变、轴向拉伸应力-应变关系、应力-应变曲线、胡克定律、比例极限、屈服点、极限应力、颈缩、脆性、延性、杨氏弹性模量以及各向同性概念。教学重点是胡克定律和应力-应变曲线，难点是应力-应变曲线。通过本单元的学习，要求熟练掌握应力应变的定义、应力-应变曲线和胡克定律。

5.1   The Concepts of Stress and Strain

Stress and strain can be illustrated in an elementary way by considering the extension of a prismatic bar. As shown in Fig. 5-1, a prismatic bar is one that has constant cross section throughout its length and a straight axis. In this illustration the bar is assumed to be loaded at its ends by axial forces P that produce a uniform stretching, or tension, of the bar. *By making an artificial cut (section m-m) through the bar at right angles to its axis, we can isolate part of the bar as a free body*[1] (Fig. 5-1b). At the left-hand end the tensile force P is applied, and at the other end there are forces representing the action of the removed portion of the bar upon the part that remains. These forces will be continuously distributed over the cross section, analogous to the continuous distribution of hydrostatic pressure over a submerged surface.

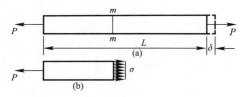

Fig. 5-1   Prismatic Bar in Tension
(a) Section *m-m*; (b) A Free Body

The intensity of force, that is, the force per unit area, is called the stress and is commonly denoted by the Greek letter σ. Assuming that the stress has a uniform distribution over the cross section (Fig. 5-1b), we can readily see that its resultant is equal to the intensity σ times the cross-sectional area A of the bar. Furthermore, from the equilibrium of the body shown in Fig. 5-1(b), we can also see that this resultant must be equal in magnitude and oppo-

site in direction to the force P. Hence, we obtain

$$\sigma = \frac{P}{A} \qquad (5\text{-}1)$$

Eq. (5-1) can be regarded as the equation for the uniform stress in a prismatic bar. This equation shows that stress has units of force divided by area. When the bar is being stretched by the force P, as shown in the figure, the resulting stress is a tensile stress; if the forces are reversed in direction, causing the bar to be compressed, they are called compressive stresses. A necessary condition for Eq. (5-1) to be valid is that the stress σ must be uniform over the cross section of the bar. This condition will be realized if the axial force P acts through the centroid of the cross section. When the load P does not act at the centroid, bending of the bar will result, and a more complicated analysis is necessary. At present, however, it is assumed that all axial forces are applied at the centroid of the cross section unless specifically stated to the contrary. Also, unless stated otherwise, it is generally assumed that the weight of the object itself is neglected, as is done when discussing the bar in Fig. 5-1.

The total elongation of a bar carrying an axial force will be denoted by the Greek letter δ (Fig. 5-1a), and the elongation per unit length, or strain, is then determined by the equation

$$\varepsilon = \frac{\delta}{L} \qquad (5\text{-}2)$$

where, L is the total length of the bar. *Note that the strain ε is a non-dimensional quantity.*[2] It can be obtained accurately from Eq. (5-2) as long as the strain is uniform throughout the length of the bar. If the bar is in tension, the strain is a tensile strain, representing an elongation or stretching of the material; if the bar is in compression, the strain is a compressive strain, which means that adjacent cross sections of the bar move closer to one another.

5.2 Axial Tensile Stress–Strain Relationship

The relationship between stress and strain in a particular material is determined by means of a tensile test. A specimen of the material, usually in the form of a round bar, is placed in a testing machine and subjected to tension. The force on the bar and the elongation of the bar are measured as the load is increased. The stress in the bar is found by dividing the force by the cross-sectional area, and the strain is found by dividing the elongation by the length along which the elongation occurs. In this manner a complete stress-strain diagram can be obtained for the material. The typical shape of the stress-strain diagram for structural steel is shown in Fig. 5-2, where the axial strains are plotted on the horizontal axis and the corresponding stresses are given by the ordinates to the curve $OABCDE$. From O to A the stress and strain are directly proportional to one another and the diagram is linear. Beyond point A the linear relationship between stress and strain no longer exists, hence the stress at A is called the proportional limit.

point B a considerable elongation begins to occur with no appreciable increase in the tensile force. This phenomenon is known as yielding of the material, and the stress at point B is called the yield point or yield stress. In the region BC the material is said to have become plastic, and the bar may actually elongate plastically by an amount which is 10 or 15 times the elongation which occurs up to the proportional limit. At point C the material begins to strain harden and to offer additional resistance to increase in load. Thus, with further elongation the stress increases, and it reaches its maximum value, or ultimate stress, at point D. Beyond this point further stretching of the bar is accompanied by a reduction in the load, and fracture of the specimen finally occurs at point E on the diagram.

During elongation of the bar a lateral contraction occurs, resulting in a decrease in the cross-sectional area of the bar. This phenomenon has no effect on the stress-strain diagram up to about point C, but beyond that point the decrease in area will have a noticeable effect upon the calculated value of stress. A pronounced necking of the bar occurs (Fig. 5-3), and if the actual cross-sectional area at the narrow part of the neck is used in calculating σ, it will be found that the true stress-strain curve follows the dashed line CE'. Whereas the total load the bar can carry does indeed diminish after the ultimate stress is reached (line DE), this reduction is due to the decrease in area and not to a loss in strength of the material itself.

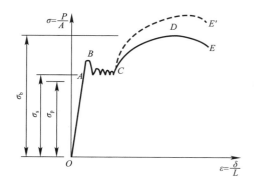

Fig. 5-2 A Typical Stress-Strain Curve

With an increase in loading, the strain increases more rapidly than the stress, until at

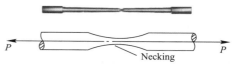

Fig. 5-3 Necking of a Tensile Bar

The material actually withstands an increase in stress up to the point of failure. For most practical purposes, however, the conventional stress-strain curve *OABCDE*, based upon the original cross-sectional area of the specimen, provides satisfactory information for design purposes. The diagram in Fig. 5-2 has been drawn to show the general characteristics of the stress-strain curve. There is an initial region on the stress-strain curve in which the material behaves both elastically and linearly. The region from *O* to *A* on the stress-strain diagram for steel is an example. The presence of a pronounced yield point followed by large plastic strains is somewhat unique to steel, which is the most common structural metal in use today. Aluminum alloys exhibit a more gradual transition from the linear to the nonlinear region. Both steel and many aluminum alloys will undergo large strains before failure and are therefore classified as ductile. On the other hand, materials that are brittle fail at relatively low values of strain. Examples include ceramics, cast iron, concrete, certain metallic alloys and glass.

5.3   Hooke's Law

5.3.1   Stress-Strain Diagram

It is apparent from this discussion that for general purposes the deformations of a rod in tension or compression are most conveniently expressed in terms of strain. Similarly, stress rather than force is the more significant parameter in the study of materials, since the effect on a material of an applied force *P* depends primarily on the cross-sectional area of the member. As a consequence, in the study of the properties of materials, it is customary to plot diagrams on which a relationship between stress and strain for a particular test is reported. Such diagrams establish a relationship between stress and strain, and for most practical purposes are assumed to be independent of the size of the specimen or its gage length. *For these stress-strain diagrams, it is customary to use the ordinate scale for stresses and the abscissa for strains.* [3] Stresses are usually computed on the basis of the original area of a specimen, although, as mentioned earlier, some transverse contraction or expansion of a material always takes place. If the stress is computed by dividing the applied force by the corresponding actual area of a specimen at the same instant, the so-called true stress is obtained. A plot of true stress vs. strain is called a true stress-strain diagram. Such diagrams are seldom used in practice.

Experimentally determined stress-strain diagrams differ considerably for different materials. Even for the same material they differ, depending on the temperature at which the test is conducted, the speed of the test and several other variables. However, broadly speaking, two types of diagrams can be recognized. One type is shown in Fig. 5-4, which is for mild steel, a ductile material widely used in construction. The other type is shown in Fig. 5-5. Such diverse materials as tool steel, concrete, copper, etc., have curves of this variety, although the extreme value of strain that these materials can withstand differs drastically. The "steepness" of these curves varies considerably. Numerically speaking, each material has its own curve. The terminal point on a stress-strain diagram represents the complete failure (rupture) of a specimen. Materials capable of withstanding large strains are referred to as ductile materials. The converse applies to brittle materials.

5.3.2   Hooke's Law

Fortunately, one feature of stress-strain diagrams is applicable with sufficient accuracy to nearly all materials. It is a fact that for a certain distance from the origin the experimental values of stress vs. strain lie essentially on a straight line. This holds true almost without reservations for glass. It is true for mild steel up to some point, as A in Fig. 5-4. It holds nearly true up to very close to the failure point for many high-grade alloy steels. On the other hand, the straight part of the curve hardly exists in concrete, annealed copper or cast iron. Nevertheless, for all practical purposes, up to some such point as A（also in Fig. 5-5）, the relationship between stress and strain may be said to be linear for all materials. This sweeping idealization and generalization applicable to all materials became known as Hooke's Law. Symbolically, this law can be expressed by the equation

$$\sigma = E\varepsilon \qquad (5\text{-}3)$$

which simply means that stress is directly proportional to strain where the constant of proportionality is E. *This constant E is called the elastic*

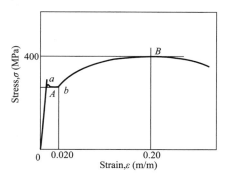

Fig. 5-4   Stress-Strain Diagram for Mild Steel

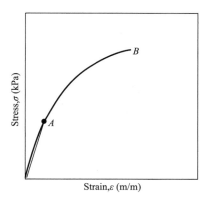

Fig. 5-5   Stress-Strain Diagram for a Brittle Material

modulus, modulus of elasticity or Young's modulus. As ε is dimensionless, E has the units of stress in this relation. In SI units it is measured in newtons per square meter（or Pascals）. [4]

Graphically E is interpreted as the slope of a straight line from the origin to the rather vague point A on a stress-strain diagram. The stress corresponding to the latter point is termed the proportional limit of the material. Physically the elastic modulus represents the stiffness of the material to an imposed load. The value of the elastic modulus is a definite property of a material. From experiments it is known that ε is always a very small quantity, hence E must be a large one. For most steels, E is between 200×10^9 and 210×10^9 N/m². It follows from the foregoing discussion that Hooke's Law applies only up to the proportional limit of the material. This is highly significant as in most of the subsequent treatment the derived formulas are based on this law. Clearly then, such formulas will be limited to the material's behavior in the lower range of stress.

Some materials, notably single crystals, possess different elastic moduli in different directions with reference to their crystallographic planes. Such materials, having different physical properties in different directions, are termed non-isotropic. A consideration of such materials is excluded from this text. The vast majority of engineering materials consist of a large number of randomly oriented crystals. Because of this random orientation of crystals, properties of materials become essentially alike in any direction. Such materials are called isotropic.

5.3.3   Future Remarks on Stress-Strain Diagrams

In addition to the proportional limit defined in Section 5.3.2, several other interesting points can be observed on the stress-strain diagrams. For instance, the highest points（B in Figs. 5-4 and 5-5）correspond to the ultimate strength of a material. Stress associated with the remarkably long plateau ab in Fig. 5-4 is termed the yield point of a material. As will be brought out later, this remarkable property of mild steel, in common with other ductile materials, is significant in stress analysis. For the present, note that at an essentially constant stress, strains 15 to 20 times those that take place up to the proportional limit occur during yielding. At the yield point a large amount of deformation takes place at a constant stress. The yielding phenomenon is absent in most materials, particularly in those that behave in a brittle fashion. A study of stress-strain diagrams shows that the yield point is so near the proportional limit that for most purposes the two may be taken as one. However, it is much easier to locate the former. For materials that do not possess a well-defined yield point, one is actually "invented" by the use of the so-called "off-set method". This is illustrated in Fig. 5-6 where a line offset an arbitrary amount of 0.2% of strain is drawn parallel to the straight-line portion of the initial stress-strain diagram. Point C is then taken as the yield point of the material at 0.2% off-set.

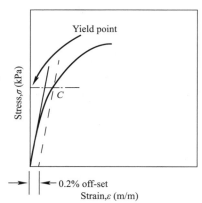

Fig. 5-6   Off-Set Method of Determining the Yield Point of a Material

Finally, the technical definition of the elasticity of a material should be made. In such usage it means that a material is able to regain completely its original dimensions upon removal of the applied forces. At the beginning of loading, if a small force is applied to a body, the body deforms a certain small amount. If such a force is removed, the body returns to its initial

size and shape. With increasing magnitude of force this continues to take place while the material behaves elastically. However, eventually a stress is reached that causes permanent deformation, or set, in the material. The corresponding stress level is called the elastic limit of the material. Practically speaking, the elastic limit corresponds closely to the proportional limit of the material. For the majority of materials, stress-strain diagrams obtained for short compression blocks are reasonably close to those found in tension. However, there are some notable exceptions. For example, cast iron and concrete are very weak in tension but not in compression. For these materials the diagrams differ considerably, depending on the sense of the applied force.

Vocabulary and Expressions：

centroid ['sentrɔid] n. 质心，形心
elasticity [elæs'tisiti] n. 弹力，弹性
elongation [ˌiːlɔŋ'geiʃən] n. 伸长，延伸
equation [i'kweiʃən] n. 方程，等式
equilibrium [ˌiːkwi'libriəm] n. 平衡
hydrostatics [ˌhaidrəu'stætiks] n. 流体静力学
linear ['liniə] adj. 线性的，直线的
modulus ['mɔdjuləs] n. 模量，模数
Young's modulus 杨氏模量
prismatic [priz'mætik] adj. 棱柱形的，棱镜的
resultant [ri'zʌltənt] adj. 合成的；n. 合力
strain [strein] n. 应变
stress [stres] n. 应力
tensile ['tensail] adj. 拉力的，张力的，拉伸的，可拉长的
alloy ['ælɔi] n. 合金
aluminum [ˌælə'miniəm] n. 铝
axial ['æksiəl] adj. 轴的，轴向的
brittle ['britl] adj. 脆性的，易碎的
contraction [kən'trækʃən] n. 收缩
diminish [di'miniʃ] v. (使)减少，(使)变小
ductile ['dʌktail] adj. 易延展的，韧性的
harden ['haːdn] vt. 使变硬，使坚强；vi. 变硬
lateral ['lætərəl] adj. 侧面的，横向的
neck [nek] n. 脖子，颈；vi. 收缩，颈缩

necking ['nekiŋ] n. 颈缩

plastic ['plæstik] adj. 塑性的

resistance [ri'zistəns] n. 抵抗力，阻力，电阻

specimen ['spesimin] n. 试件

yield [ji:ld] n. 屈服；v. 产出，生产；vi. (~ to) 屈服于，屈从；~ point 屈服点，屈服强度

Hooke's Law 胡克定律

stress-strain diagram 应力-应变图

steepness ['sti:pnis] n. 倾斜度

annealed copper 退火铜，韧铜

dimensionless [di'menʃənlis] adj. 无量纲的，没有单位的

crystallographic [ˌkristələu'græfik] adj. 晶体学的，结晶的，结晶学的

isotropic [ˌaisəu'trɔpik] adj. 各向同性的

Notes：

① 通过用假想截面 m-m 沿与杆件轴线垂直的方向将杆件切开，我们可以将杆体的一部分分离成为一个自由脱离体。

② 注意应变 ε 是一个无量纲的量。

③ 在应力-应变图中，习惯将应力作为纵坐标，而将应变作为横坐标。

④ 常数 E 叫做弹性模量或杨氏模量。由于 ε 是无量纲的，所以 E 的单位和应力一样，在国际标准单位里就是 N/m^2（或 Pa）。SI 是指 Standard International Units 国际标准单位。

Exercises：

1. Translate the Following Paragraphs into Chinese.

（1）Noe that E has the same units as stress. The modulus of elasticity is sometimes called Young's modulus, after the English scientist Thomas Young (1773 ~ 1829) who studied the elastic behavior of bars. For most materials the modulus of elasticity in compression is the same as in tension.

（2）A study of stress-strain diagrams shows that the yield point is so near the proportional limit that for most purposes the two may be taken as one. However, it is much easier to locate the former. For materials that do not possess a well-defined yield point, one is actually "invented" by the use of the so-called "off-set method". This is illustrated in Fig. 5-6 where a line offset an arbitrary amount of 0. 2% of strain is drawn parallel to the straight-line portion of the initial stress-strain diagram. Point C is then taken as the yield point of the material at 0. 2% off-set.

2. Translate the Following Paragraphs into English.

（1）平面假设是指变形前原是平面的截面在变形后仍保持为平面。

（2）大多数建筑材料在受力不超过弹性范围时，其横截面上正应力和轴向线应变成正比。材料受力后其应力和应变之间的这种比例关系称为胡克定律。

（3）构件在工作时允许承受的最大工作应力，称之为许用应力，以符号 $[\sigma]$ 表示。许用应力等于极限应力 σ_u 除以安全系数 n，即 $[\sigma] = \sigma_u / n$。

6 Differential Equation for Beam Deflection Curve and Moment-Area Method

（梁的挠曲线微分方程及弯矩面积法）

教学目标：本单元的主要内容有简支梁、梁的挠曲线方程、曲率、曲率半径、挠度、积分常数、边界条件、弯矩面积法、惯性矩、剪力图、弯矩图、静定梁、静力平衡方程、超静定梁以及变形协调方程。教学重点是变形协调方程，难点是梁的挠曲线方程。通过本单元的学习，要求熟悉惯性矩、剪力图、弯矩图、边界条件、静定梁和超静定梁概念，掌握利用静力平衡方程和变形协调方程求解超静定梁的内力，掌握使用梁的挠曲线方程计算梁任一截面的挠度。

6.1   Deflection Curve of a Bent Beam

A bar that is subjected to forces acting transverse to its axis is called a beam. *The beam in Fig.* 6-1, *with a pin support at one end and a roller support at the other*, *is called a simply supported beam or a simple beam.* [①] The essential feature of a simple beam is that both ends of the beam may rotate freely during bending, but they cannot translate in the lateral direction. Also, one end of the beam can move freely in the axial direction. The beam, which is built-in or fixed at one end and free at the other end, is called a cantilever beam. At the fixed support the beam can neither rotate nor translate, while at the free end it may do both.

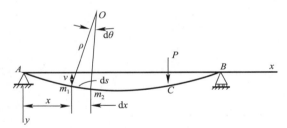

Fig. 6-1   Deflection Curve of a Bent Beam

6.2   Differential Equation for Beam Deflection Curve

Loads on a beam may be classified into concentrated force, such as P in Fig. 6-1, or distributed loads which is expressed in units of force per unit distance along the axis of the beam. The axial force N acting normal to the cross section and passing through the centroid of the cross section, shear force V acting parallel to the cross section and bending moment M acting in the plane of the beam are known as stress resultants. The relationship between the shear force V, bending moment M and the loads on a beam is given by

$$\frac{\mathrm{d}M}{\mathrm{d}x} = V \qquad (6\text{-}1)$$

This equation shows that the rate of change of the bending moment is equal to the algebraic value of the shear force, provided that a distributed load (or no load) acts on the beam. If the beam is acted upon by a concentrated force, however, there will be an abrupt change, or discontinuity, in the shear force at the point of application of the concentrated force. Lateral loads acting on a beam will cause the beam to deflect. As shown in Fig. 6-1, before the load P

is applied, the longitudinal axis of the beam is straight. After bending, the axis of the beam becomes a curve, as represented by the line ACB. Let us assume that the xy-plane is a plane of symmetry of the beam and that all loads act in this plane. Then the curve ACB, called the deflection curve of the beam, will lie in this plane also.

From the geometry of the figure we see that

$$K = \frac{1}{\rho} = \frac{\mathrm{d}\theta}{\mathrm{d}s} \qquad (6\text{-}2)$$

where, K is the curvature, equal to the reciprocal of the radius of curvature ρ. Thus, the curvature K is equal to the rate of change of the angle θ with respect to the distance s measured along the deflection curve. The basic differential equation for the deflection curve of a beam is given as follows

$$\frac{\mathrm{d}^2 v}{\mathrm{d}x^2} = -\frac{M}{EI} \qquad (6\text{-}3)$$

where, v is the deflection of the beam from its initial position. It must be integrated in each particular case to find the deflection v. The procedure consists of successive integration of the equations, with the resulting constants of integration being evaluated from the boundary conditions of the beam. It should be realized that Eq. (6-3) is valid only when Hooke's Law applies for the material and when the slopes of the deflection curve are very small.

6.3　The Moment-Area Method

Another method for finding deflections of beams is the moment-area method. [2] The name of this method comes from the fact that it utilizes the area of the bending moment diagram. This method is especially suitable when it is desired to find the deflection or slope at only one point of the beam rather than finding the complete equation of the deflection curve. The normal and shear stresses acting at any point in the cross section of a beam can be obtained by using the equations

$$\sigma_x = \frac{My}{I}, \qquad \tau = \frac{VQ}{Ib} \qquad (6\text{-}4)$$

in which I is the second moment (or the moment of inertia) of the cross-sectional area with respect to the neutral axis; Q is the first moment (or static moment) of the plane area of a beam. It can be seen that the normal stress is a maximum at the outer edges of the beam and is zero at the neutral axis; the shear stress is zero at the outer edges and usually reaches a maximum at the neutral axis.

The shear force V and bending moment M in a beam will usually vary with the distance x defining the location of the cross section at which they occur. When designing a beam, it is desirable to know the values of V and M at all cross sections of the beam, and a convenient way to provide this information is by a graph showing how they vary along the axis of the beam. *To plot the graph, we take the abscissa as the position of the cross section, and we take the ordinate as the corresponding value of either the shear force or the bending moment. Such graphs are called shear force and bending moment diagrams.* [3]

The simple beam shown in Fig. 6-1 is one of the statically determinate beams. The feature of this type of beams is that all their reactions can be determined from equations of static equilibrium. The beams that have a large number of reactions than the number of equations of static equilibrium are said to be statically indeterminate. For statically determinate beams we could immediately obtain the reactions of the beam by solving equations of static equilibrium. However, when the beam is statically indeterminate, we cannot solve for the forces on the basis of statics alone. Instead, we must take into account the deflections of the beam and obtain equations of compatibility to supplement the equations of statics.

Vocabulary and Expressions：

translate ［trænz'leit，træns'leit］ v. 翻译，转化，转变为，变换，转移，中继，移动，［力学］平移

boundary ［'baundəri］ n. 边界，分界线

cantilever ［'kæntiliːvə］ n. 悬臂（梁），伸臂

compatibility ［kəmˌpætə'biliti］ n. 相容性，协调性

concentrate ［'kɔnsentreit］ v. 集中，浓缩，专心

curvature ［'kəːvətʃə］ n. 曲率，弯曲

deflection ［di'flekʃən］ n. 挠度，挠曲；偏转，偏斜

determinate ［di'təːminit］ adj. 确定的，决定的；n. ［数］行列式

differential ［ˌdifə'renfəl］ adj. 微分的；n. 微分

discontinuity ［'disˌkɔnti'nju(ː)iti］ n. 间断，不连续

distribute ［dis'tribu(ː)t］ vt. 分布，散布，分配

indeterminate ［ˌindi'təːminit］ adj. 不确定的

integrate ［'intigreit］ vt. 使成整体，使一体化，求…的积分；v. 结合，综合，集成

normal ［'nɔːməl］ adj. 正常的，正规的，法向的

reciprocal ［ri'siprəkəl］ adj. 倒数的，彼此相反的

slope ［sləup］ n. 斜率，倾斜

statics ［'steitiks］ n. 静力学，静止状态，静态

successive ［sək'sesiv］ adj. 连续的，相继的

symmetry ［'simitri］ n. 对称

transverse ［'trænzvəːs］ adj. 横向的，横断的

Notes：

① 图 6-1 中一端有固定铰支座支撑，另一端有可动铰支座支撑的梁，称为简支梁。

② 另一种求梁挠度的方法是弯矩面积法。

③ 为了画出这幅图，我们取横坐标作为截面的位置，纵坐标作为相应的剪力或弯矩的值。这些图称为剪力图和弯矩图。

Exercises：

1. Translate the Following Paragraph into Chinese.

The simple beam shown is one of the statically determinate beams. The feature of this type of beams is that all their reactions can be determined from equations of static equilibrium. The beams that have a large number of reactions than the number of equations of static equilibrium are said to be statically indeterminate. For statically determinate beams we could immediately obtain the reactions of the beam by solving equations of static equilibrium. However, when the beam is statically indeterminate, we cannot solve for the forces on the basis of statics alone. Instead, we must take into account the deflections of the beam and obtain equations of compatibility to supplement the equations of statics.

2. Translate the Following Paragraphs into English.

（1）只要知道梁的挠曲线方程，任意截面的位移、转角都可以确定。

（2）在坐标系中，规定向下的挠度为正，向上的挠度为负；顺时针转角为正，逆时针转角为负。

（3）将上式两边积分一次，可得转角方程，积分两次可得挠曲线方程，积分常数由梁变形的连续条件及边界条件确定。

（4）梁的变形不仅取决于挠曲线的曲率，还与支座的约束条件有关。

（5）叠加法计算梁的变形：在小变形、材料服从胡克定律的条件下，当梁上有几个荷载共同作用时，梁横截面的挠度或转角，等于单个荷载单独作用时引起的该截面的挠度或转角的代数和。

7 Torsion of a Circular Bar and Hook' s Law for Shear
（圆杆扭转与剪切胡克定律）

教学目标：本单元的主要内容有圆杆扭转、转角、剪应力、剪应变、剪切胡克定律、扭矩计算公式、极惯性矩以及抗扭刚度。教学重点是剪切胡克定律，难点是扭矩计算公式。通过本单元的学习，要求理解剪应力、剪应变、极惯性矩和抗扭刚度的概念，掌握剪切胡克定律，了解扭矩计算公式的推导。

7.1　Circular Bar in Pure Torsion

Let us consider a bar of circular cross section twisted by couples T acting at the ends (Fig. 7-1). A bar loaded in this manner is said to be in pure torsion. It can be shown from considerations of symmetry that cross sections of the circular bar rotate as rigid bodies about the longitudinal axis, with radii remaining straight and the cross sections remaining circular. Also, if the total angle of twist of the bar is small, neither the length of the bar nor its radius r will change.

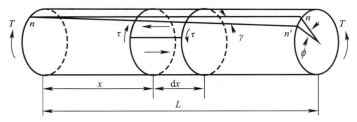

Fig. 7-1　Circular Bar in Pure Torsion

7.2　Hook's Law for Shear

During torsion there will be a rotation about the longitudinal axis of one end of the bar with respect to the other. For instance, if we consider the left-hand end of the bar as fixed, then the right-hand end will rotate through an angle ϕ with respect to the left end. At the same time, a longitudinal line on the surface of the bar, such as line nn, will rotate through a small angle to the position nn'. Because of this rotation, a rectangular element on the surface of the bar, such as the element shown in the figure between two cross sections distance dx apart, is distorted into a rhomboid. When a shaft is subjected to pure torsion, the rate of change $d\phi/dx$ of the angle of twist is constant along the length of the bar. This constant represents the angle of twist per unit length and will be denoted by θ. Thus,

we see that $\theta = \phi/L$, where L is the length of the shaft. Then, the shear strain γ is obtained by $\gamma L = nn' = r\phi = r\theta L$, i.e.

$$\gamma = r\theta = r\phi/L \qquad (7\text{-}1)$$

The shear stresses τ which act on the sides of the element have the directions shown in Fig. 7-1.

For a linear elastic material, the magnitude of the shear stress is

$$\tau = G\gamma = Gr\theta \qquad (7\text{-}2)$$

Eqs. (7-1) and (7-2) relate the strain and stress at the surface of the shaft to the angle of twist per unit length. Eq. (7-2) is called Hook's Law for Shear.

The state of stress within the interior of the shaft can be determined in a manner similar to that used for the surface of the shaft. Because radii in the cross sections of the bar remain straight and undistorted during twisting, we see that an interior element situated on the surface of an interior cylinder of radius ρ is also in pure shear with the corresponding shear strain and stress being given by the following expressions

$$\gamma = \rho\theta \qquad (7\text{-}3a)$$
$$\tau = G\rho\theta \qquad (7\text{-}3b)$$

These equations show that the shear strain and shear stress vary linearly with the radial distance ρ from the center of the shaft and have their maximum values at the outer surface.

The shear stresses acting in the plane of the cross section, given by Eq. (7-3b), are accompanied by equal shear stresses acting on longitudinal planes of the shaft. This result follows from the fact that equal shear stresses always exist on mutually perpendicular planes. *If a material is weaker in shear longitudinally than laterally (for example, wood), the first cracks in a twisted shaft will appear on the surface in the longitudinal direction.* [1] *The state of pure shear stress on the surface of the shaft is equivalent to the tensile and compressive stresses on an element rotated through an angle of 45° to the axis of the shaft. If a material that is weaker in tension than in shear is twisted, failure occurs in tension along a helix inclined at 45° the axis.* [2] This type of failure can easily be demonstrated by twisting a piece of chalk.

7.3　Computational Formula of T

The relationship between the applied torque T and the angle of twist which it produces will now be established. The resultant of the shear stresses must be statically equivalent to the total torque T. The shear force acting on an element of area dA is τdA, and the moment of this force about the axis of the bar is $\tau\rho d A$. Using Eq. (7-3b), this moment is also equal to $G\theta\rho^2 dA$. The total torque T is the summation over the entire cross-sectional area of such elemental moments, thus

$$T = \int G\theta\rho^2 dA = G\theta\int\rho^2 dA = G\theta I_P$$
$$(7\text{-}4)$$

where, $I_P = \int\rho^2 dA$ is the *polar moment of inertia* [3] of the circular cross section. From Eq. (7-4) we obtain $\theta = T/GI_P$ which shows that θ, the angle of twist per unit length, varies directly with the torque T and inversely with the product GI_P, known as the *torsional rigidity* [4] of the shaft.

Vocabulary and Expressions:

couple ['kʌpl] n. 对，力偶；v. 耦连，耦合
crack [kræk] n. 裂缝，裂纹；v. （使）破裂
cylinder ['silində] n. 圆筒，圆筒状物
distort [dis'tɔ:t] vt. 扭曲，弄歪
helix ['hi:liks] n. 螺旋，螺旋状物

inertia [i'nə:ʃjə] n. 惯性，惯量，惰性
longitudinal [lɔndʒi'tju:dinl] adj. 纵向的，经线的
moment ['məumənt] n. 力矩，瞬间；adj. 片刻的，瞬间的，力矩的
polar ['pəulə] adj. 极性的，两极的
radius ['reidjəs] n. 半径
rhomboid ['rɔmbɔid] n. 长菱形
rigidity [ri'dʒiditi] n. 刚性，刚度，坚硬，僵硬
shaft [ʃa:ft] n. 轴，杆状物
torque [tɔ:k] n. 扭矩
torsion ['tɔ:ʃən] n. 扭转，扭曲
twist [twist] vt. 扭，拧，使扭转；n. 扭曲

Notes:

① 如果一种材料的纵向剪切比横向剪切弱（如木材），扭转轴的第一批裂缝将出现在纵向表面上。
② 轴表面的纯剪应力状态与轴旋转45°角后的单元体上的拉压应力状态等效。如果使用一种抗拉强度比抗剪强度弱的材料进行扭转，将发生在和轴呈45°倾斜的螺旋线上的拉应力破坏。
③ 极惯性矩。
④ 抗扭刚度。

Exercises:

1. Translate the Following Paragraph into Chinese.

When a shaft is subjected to pure torsion, the rate of change $d\phi/dx$ of the angle of twist is constant along the length of the bar. This constant represents the angle of twist per unit length and will be denoted by θ. Thus, we see that $\theta = \phi/L$, where L is the length of the shaft. Then, the shear strain γ is obtained by $\gamma L = nn' = r\phi = r\theta L$, i.e.

$$\gamma = r\theta = r\phi/L \qquad (7\text{-}1)$$

The shear stresses τ which act on the sides of the element have the directions shown in Fig. 7-1.

For a linear elastic material, the magnitude of the shear stress is

$$\tau = G\gamma = Gr\theta \qquad (7\text{-}2)$$

Eqs. (7-1) and (7-2) relate the strain and stress at the surface of the shaft to the angle of twist per unit length.

2. Translate the Following Paragraphs into English.

（1）扭矩正负号规定如下：以拇指代表横截面的外法线方向，则与其余四指的转向相同的扭矩为正，反之为负。

（2）扭矩图的横坐标与轴线平行，代表横截面的位置，纵坐标代表相应截面上的扭矩值，正扭矩画在横坐标的上方，负扭矩画在横坐标的下方，从扭矩图上可确定最大扭矩值及其所在横截面的位置。

（3）剪应力互等定理：两个互相垂直平面上的切应力大小相等，其方向同时指向（或背离）两个平面的交线。

（4）剪切胡克定律：在线弹性范围内，切应力与切应变成正比。

8 Critical Load and Euler's Formula
（临界荷载与欧拉公式）

教学目标：本单元的主要内容有临界力（或临界荷载）及其计算公式推导、压杆稳定、欧拉公式、柔度（或长细比）以及临界应力。教学重点是欧拉公式和临界应力，难点是临界力计算公式的推导。通过本单元的学习，要求理解压杆稳定、柔度（或长细比）以及临界应力的概念，掌握欧拉公式和临界应力公式的应用计算。

8.1　Critical Load

The selection of the columns is often a very crucial part of the design of a structure because the failure of a column usually has catastrophic effects. Furthermore, columns are more difficult to design than bars in bending or torsion because their behavior is more complicated. If a column which is long compared to its width is subjected to axial force F, it may fail by buckling, that is, the deflection increases rapidly as the load F approaches a certain critical value. This value is known as the *critical load* [1] F_{cr}. The buckling phenomenon is associated with the transition of the column configuration from a stable equilibrium condition to an unstable equilibrium condition when the critical load F_{cr} is reached. In order to investigate the behavior of columns, we will begin by considering a slender, perfectly straight column of length L which is fixed at the lower end and free at the upper end (Fig. 8-1a). If the axial load F is less than the critical value, the bar remains straight and undergoes only axial compression. This straight form of equilibrium is stable, which means that if a lateral force is applied and a small deflection is produced, the deflection will disappear, and the bar will return to its straight form, when the lateral force is removed.

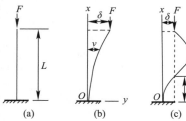

Fig. 8-1　A Slender Column
(a) A Slender Column before Buckling; (b) Buckled Mode Shape for $n=1$;
(c) Mode Shape for $n=3$; (d) Mode Shape for $n=5$

8.2　Critical Load Formula

However, as F is gradually increased, a condition of neutral equilibrium is reached when F becomes equal to F_{cr}. At this load the column theoretically may have any small value of deflection, and a small lateral force will produce a deflection which does not disappear when the lateral force is removed. At higher values of the load the column is unstable and will collapse. The critical load for a column can be calculated by using the equation of the deflection curve. For the column shown in Fig. 8-1(b), the equation is

$$EIv'' = F(\delta - v) \qquad (8-1)$$

where δ is the deflection at the free end. Using the notation $k^2 = F/EI$, we can write the general solution of Eq. (8-1) in the form

$$v = C_1\sin kx + C_2\cos kx + \delta \qquad (8-2)$$

The boundary conditions at the built-in end of the column, $v = v' = 0$ gives $C_1 = 0$ and $C_2 = -\delta$, and the deflection curve becomes $v = \delta(1 - \cos kx)$. Using the boundary condition at the upper end of the column, $v = \delta$, we find

$$\delta \cos kL = 0 \qquad (8\text{-}3)$$

from which we conclude that either $\delta = 0$ or $\cos kL = 0$. If $\delta = 0$, there is no deflection of the column and hence no buckling. Such an event is represented in Fig. 8-1(a). The other possibility is that $\cos kL = 0$, we see from Eq. (8-3) that δ can have any small value. The condition $\cos kL = 0$ requires that $kL = n\pi/2$ where $n = 1$, 3, 5, $\cdots$ Thus, we can obtain an infinite number of critical loads

$$F_{cr} = \frac{n^2 \pi^2 EI}{4L^2} \qquad (8\text{-}4)$$

This equation shows that, as n is increased, the deflection curve has more and more waves in it. When $n = 1$, the curve has one-half of a wave, as shown in Fig. 8-1(b). The deflection curves for $n = 3$ and $n = 5$ are shown in Figs. 8-1(c) and (d) respectively. Although they represent theoretically possible modes of buckling for the column, they are of no practical interest because the column will buckle in the first mode which corresponds to the smallest critical load of the column.

Critical buckling loads for columns with simply supported ends and with fixed ends can be obtained from the solution of the preceding case. For example, it is evident from symmetry that the deflection curve of a column with hinged ends in the first mode of buckling will have a vertical tangent at the midpoint. Hence, each half of the column is in the same condition as the bar in Fig. 8-1(b), and the critical load is obtained from Eq. (8-4) ($n = 1$) by substituting $L/2$ for L, i.e., $F_{cr} = \pi^2 EI/L^2$. If the column is fixed at both ends, the deflection curve for the first buckling mode is a cosine curve having inflection points at distance $L/4$ from the ends. Therefore, the critical load is obtained by substituting $L/4$ for L, i.e., $F_{cr} = 4\pi^2 EI/L^2$.

From Eq. (8-4) we see that the critical load for a column is directly proportional to the flexural rigidity EI and inversely proportional to the square of the length. [2] Thus, the critical load can be increased by increasing the moment of inertia I of its cross section. This result can be accomplished by distributing the material as far as possible from the centroid of the cross section. *Hence, tubular members are more economical for columns than are solid members having the same cross-sectional area.* [3] By diminishing the wall thickness of such sections and increasing the transverse dimension, the stability of the column is increased because I is greater. There is a lower limit for the wall thickness, however, below which the wall itself becomes unstable. Then, instead of buckling of the column as a whole, there will be localized buckling of the wall in the form of corrugations of wrinkling of the wall. This type of buckling is called local buckling and requires a more detailed investigation.

8.3　Euler's Formula

8.3.1　Column Buckling

The transition from stable equilibrium to unstable equilibrium in the straight line state of the pressure bar is called the instability of the pressure bar. The critical force F_{cr} is an important index to judge whether the pressure bar will lose stability. When $F < F_{cr}$, the equilibrium is stable; when $F \geqslant F_{cr}$, the equilibrium is unstable.

8.3.2　Euler's Formula

The critical force of compression bar with different rod end constraints can be uniformly written into Euler's Formula as follows

$$F_{cr} = \frac{\pi^2 EI}{(\mu l)^2}$$

where, μ is the length coefficient.

8.4   Critical Stress and Flexibility (or Slenderness Ratio)

The stress obtained by dividing the critical force by the cross section area of the compression bar is called the critical stress and is expressed as σ_{cr}

$$\sigma_{cr} = \frac{F_{cr}}{A} = \frac{\pi^2 E}{\lambda^2}$$

In the formula, λ ($\lambda = \frac{\mu l}{i}$) is called the flexibility or slenderness ratio. The greater the flexibility of the pressure bar is, the more easily it is to lose stability.

Vocabulary and Expressions：

buckle ['bʌkl] n. 屈曲，皱曲；v. 弄弯，皱曲，翘曲

built-in ['bilt'in] adj. 固定的，嵌入的

catastrophic [,kætə'strɔfik] adj. 灾难的

collapse [kə'læps] n., vi. 倒塌，崩溃

column ['kɔləm] n. 圆柱，列

compression [kəm'preʃən] n. 压缩

corrugation [,kɔru'geiʃən] n. 起皱，波纹

crucial ['kruːʃiəl, 'kruːʃəl] adj. 决定性的

deflection [di'flekʃən] n. 挠曲，偏向

flexural ['flekʃərəl] adj. 弯曲的，挠曲的

hinge [hindʒ] n. 铰链，枢纽；v. 以…而定；以…为转移

infinite ['infinit] adj. 无穷的，无限的；n. 无穷大

inflection [in'flekʃən] n. 弯曲，屈曲

moment ['məumənt] adj. 片刻的，瞬间的，力矩的；n. 瞬间，矩，力矩

notation [nəu'teiʃən] n. 记号，注释

slender ['slendə] adj. 细长的，苗条的

tangent ['tændʒənt] adj. 切线的，相切的；n. 切线，[数] 正切

transition [træn'ziʃn, træn'siʃn] n. 转变，过渡

tubular ['tjuːbjulə] adj. 管状的

wave [weiv] n. 波动，波浪

wrinkle ['rinkl] n. 皱纹，皱褶；v. 起皱

Notes：

① 临界荷载。

② 由式(8-4)可知，柱的临界荷载与抗弯刚度 EI 成正比，与长度的平方成反比。

③ 因此，对于柱而言，管状构件比具有相同截面面积的实心构件更经济。

Exercises：

1. Translate the Following Paragraph into Chinese.

By diminishing the wall thickness of such sections and increasing the transverse dimension, the stability of the column is increased because I is greater. There is a lower limit for the wall thickness, however, below which the wall itself becomes unstable.

2. Translate the Following Paragraphs into English.

(1) 所谓稳定性，这里是指平衡状态的稳定性。压杆直线状态的平衡由稳定平衡过渡到不稳定平衡称压杆失去稳定，简称失稳。临界力 F_{cr} 是判别压杆是否会失稳的重要指标。当 $F < F_{cr}$ 时，平衡是稳定的；$F \geqslant F_{cr}$ 时，平衡是不稳定的。不同杆端约束的压杆临界力可统一写成如下的欧拉公式

$$F_{cr} = \frac{\pi^2 EI}{(\mu l)^2}$$

式中 μ 为长度系数。

(2) 将临界力除以压杆的横截面面积，所得的应力称为临界应力，用 σ_{cr} 表示。欧拉公式表明：临界力 F_{cr} 与杆抗弯刚度 EI 成正比，与杆长 l 的平方成反比。压杆的柔度越大，越容易失稳。可以用增强杆端约束的办法减少长度系数 μ 的值，以达到降低柔度，从而提高压杆稳定性的目的。

9　Civil and Building Engineering
（土木与建筑工程）

教学目标： 本单元的主要内容有土木工程、土木工程师、建筑工程特点、建筑产品特点、建筑施工特点、建筑项目管理特点、建筑功能、围护结构、承载结构、地基、墙、屋楼盖、楼梯、门窗、阳台以及建筑分类。教学重点是建筑功能，难点是承载结构。通过本单元的学习，要求熟悉建筑物的各组成部分和建筑工程特点，理解建筑功能、围护结构和承载结构，了解建筑分类。

9.1　Definition and Scope of Civil Engineering

The word civil derives from the Latin for citizen. In 1782, Englishman John Seaton used the term to differentiate his nonmilitary engineering work from that of the military engineers who predominated at the time. Civil engineering is the planning, design, construction and management of the built environment. This environment includes all structures built according to scientific principles, from irrigation and drainage systems to rocket launching facilities.

Civil engineers must make use of many different branches of knowledge including mathematics, theory of structures, hydraulics, soil mechanics, surveying, hydrology, geology and economics. Civil engineering was not distinguished from other branches of engineering until 200 years ago. Most early engineers were engaged in the construction of fortifications and were responsible for building the roads and bridges required for the movement of troops and supplies. The Roman armies of occupation in Europe were served by brilliant engineers but after the collapse of the Roman Empire there was little progress in communications until the beginning of the Industrial Revolution, the invention of the steam engine and the realization of the potentialities in the use of iron. Roads, canals, railways, ports, harbors and bridges are then built by engineers who adopted the prefix "civil" to distinguish them from the military engineers and to emphasize the value of their work to the community.

Because it is so broad, civil engineering is subdivided into a number of technical specialties. Depending on the type of project, the skills of many kinds of civil engineer specialists may be needed. When a project begins, the site is surveyed and mapped by civil engineers who experiment to determine if the earth can bear the weight of the project. Environmental specialists study the project's impact on the local area, the potential for air and groundwater pollution, the project's impact on local animal and plant life, and how the project can be designed to meet government requirements aimed at protecting the environment. Transportation specialists determine what kinds of facilities are needed to ease the burden on local roads and other transportation networks that will result from the completed project. Meanwhile, structural specialists raise preliminary data to make detailed designs, plans and specifications for the project. Supervising and coordinating these civil engineer specialists from beginning to end of the project are the construction management specialists. Based on information supplied by the other specialists, construction management civil engineers estimate quantities and costs of materials and labor, schedule all work, order materials and equipment for the job, hire contractors and subcontractors, and perform other supervisory work to ensure the project is completed on time and as specified.

Wood, brick and stone were generally used in constructing bridges until the 19th century. Many stone bridges in the United Kingdom are several hundred years old. The first iron bridge was completed in 1779. Iron is commonly used for bridge construction for more than one hundred years. Iron has been replaced by steel, which is stronger, and permits lighter members to be used. Bridges are now made of steel or of reinforced concrete, or of steel and reinforced concrete combined, or even of aluminum alloy. Many modern bridges are made of concrete beams supported by concrete piers and abutments. Concrete is strong in compression, but weak in tension. Steel, which is strong in tension, is used to counteract the tensile forces in loaded beams. For abutments and piers and also beams, an arrangement of steel wires or rods is encased in concrete, to make reinforced concrete. When a beam is supported at its ends, the upper part of the beam is in compression and the lower part is in tension. When the beam is subjected to a heavy load the stresses are increased. If a beam is made of concrete it is usually strengthened or reinforced by means of steel rods in the lower part of the beam where the tensile stresses are found.

Suspension bridges are generally used for very long spans. A pair of cables, anchored at both ends, is supported on towers. The deck is joined to the cables by suspenders. [①] The cables are in tension, and are usually made of many thousands of fine steel wires, which together have a greater tensile strength than ordinary steel. The deck is generally made of reinforced concrete or steel. The design of suspension bridges continues to be improved. In a bridge recently built the cross-section of the deck has been streamlined to minimize the effects of cross-winds. The hollow steel boxes of which this deck is constructed require less steel than a solid deck of the same strength. The lighter deck can be supported by lighter cables and suspenders. Thus, the new design economizes on the use of steel.

All dams need sound foundations, and the nature of the bedrock under many dams entails extensive grouting of fissures and porous zones with cement or with chemicals to prevent water passing under or round the dam. Spillways which can safely pass floodwater are of many types, and often require the boring of large tunnels. Tunnels are also used to supply water to power stations, and may be combined with the steel penstocks which are often the only visible evidence of hydroelectric power schemes. It is vital to remember that a dam is designed to stop the passage of water and to store it. Without means of controlling and passing both normal and flood flows round or through the dam, the safety of the dam and of the people living in its shadow and further downstream could not be assured.

The civil engineer who chooses a teaching career usually teaches both graduate and undergraduate students in technical specialties. Many teaching civil engineers engage in basic research that eventually leads to technical innovations in construction materials and methods. Many also serve as consultants on engineering projects, or on technical boards and commissions associated with major projects.

Building engineering is a specialized branch of civil engineering concerned with the planning, execution and control of construction operations for such projects as highways, buildings, dams, airports and utility lines. Planning consists of scheduling the work to be done and selecting the most suitable construction methods and equipment for the project. Execution requires the timely mobilization of all drawings, layouts and materials on the job to prevent delays to the work. Control consists of analyzing progress and cost to ensure that the project will be done on schedule and within the estimated cost.

Construction operations are generally classified according to specialized fields. These include preparation of the project site, earthmoving, foundation treatment, steel erection, concrete placement, asphalt paving and electrical and mechanical installations.

9.2　Characteristics of Building Engineering

Building Engineering involves a subject of investigating the design, construction and repair of all kinds of buildings through applications of knowledge and techniques of various subjects such as architecture, geology, surveying, soil mechanics, engineering mechanics, building materials, building structures, construction machinery and so on. Compared with other industrial products, the architectural product has a series of its own special features that can be presented in the product itself, its construction and management.

9.2.1 Characteristics of Product

(1) Any architectural product is located in an appointed place and stands on the ground so that it is unmovable. Since different places have different conditions of topography and geology, it is necessary to design its ground base and foundation for each building. (2) Architectural products should be versatile to meet various needs of users, including utilization function, scale, structural form, style, comfort and economy. (3) Architectural products are large in size horizontally or vertically or both. (4) A building is subjected to many types of loading conditions. Besides, it is concerned with the art style, architectural function, structural construction, building material, decorative process, etc. A lot of complicated problems would be encountered during the construction of architectural products. They are not always solved theoretically, sometimes by means of experiments and experiences in fact.

9.2.2 Characteristics of Construction

(1) Building construction needs a great deal of labor, building materials and machinery

due to large-sized architectural products and complicated techniques involved. For this reason, it usually takes a long period of time to complete a building, for example, from months to years. (2) Industrial products are manufactured in the fixed workshops of factories, whereas architectural products are often provided at different locations or construction sites, from place to place. (3) The construction of buildings is performed in the open air, extremely easy to be influenced by the conditions of nature and environment.

Construction is a complicated process on almost all engineering projects. It involves scheduling the work and utilizing the equipment and the materials so that costs are kept as low as possible. Safety factor must also be taken into account, since construction can be very dangerous. Many civil engineers therefore specialize in the construction phase.

9.2.3 Characteristics of Project Management

(1) A building product should be taken as a project to be performed. For the project, there are definite objectives, detailed tasks, specified construction progresses, estimate costs, quality standards and so on. It is very important to work out a detailed scheme of managing the project. (2) Project management in building engineering involves multiple objectives, concerned with quality, construction period and budget. Many construction departments and specialist professions are involved in a project that often has to be finished in several phases. So a lot of uncertain events can potentially influence the management progress of a project.

9.3 Composition and Function of Building

A building is composed of many elements. For example, take the house in Fig. 9-1, which

includes foundations, walls or columns, floors, roots, staircases, doors, windows, balconies, etc. They can be classified into two parts: en- closing structures and load-bearing structures.

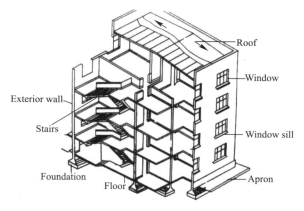

Fig. 9-1 Composition of a Dwelling Building

9.3.1 Enclosing Structures

The function of a building is achieved by means of enclosing structures that make the building enclosed and stop wind, rain and snow in, but let sunlight and fresh air into the building, allowing people to live or work both safely and comfortably. Walls, doors, windows, eaves, parapets and canopies belong to the enclosing structures.

9.3.2 Load-Bearing Structures

As shown in Fig. 9-2, a building is subjected to the weight from itself and the force from the nature such as the weights of people, furniture, equipment and structural materials, snow, wind, earthquake force and so on, which are all termed loads. The structures that are able to bear these loads and transfer them to other parts in the building are referred to load-bearing structures. The system which is composed of load-bearing structures is called structural system, which comprises a variety of structural members such as beams, slabs, columns, roof trusses, bearing walls, foundations, etc. Loads on a building are transferred from top to bottom.

(1) Foundation: The foundation is the lowest part of a building, whereas the soil under the foundation is termed ground base that does not belong to the part of the building. The foundation sustains all loads of the building and then transfers them to the ground base.

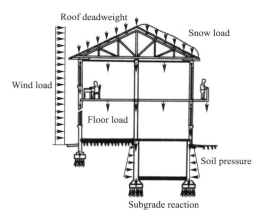

Fig. 9-2 Loads on a Building

(2) Walls: Walls can be divided into exterior walls and interior walls based on their positions, as shown in Fig. 9-3. The exterior walls are outside enclosing structures that insulate rooms in the building against wind, rain, heat and sound. The function of the interior walls is to divide the space in each floor into some individual rooms, besides heat and sound insulation. *The walls arranged along the brachy-axis and the macro-axis of a building are called transverse (or cross) and longitudinal walls respectively. The exterior cross wall is usually termed a gable wall.* [2] Walls can also be classi-

fied into bearing walls and non-bearing walls according to different structural behavior. A wall used to support any load from the roof or the upper floor is called the bearing wall. Other walls are called non-bearing walls. Walls such as partition walls, filler walls and curtain walls are examples of non-bearing walls. The walls used to fill the areas between columns of frame structures are referred to as filler walls. Curtain walls are any kind of lightweight exterior walls, i.e., glass curtain walls, which are suspended on skeleton structures or floor slabs. Walls can be made of different materials like brick, stone, earth, concrete and so on.

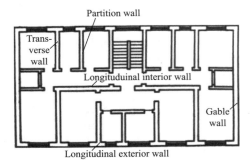

Fig. 9-3   Description of Walls

(3) Floor: A floor is a structural member arranged horizontally, which separates the space between two storeys. It is required that the floor not only can carry both vertical and horizontal loads, but also can provide sound insulation, fireproofing and waterproofing. Space also needs

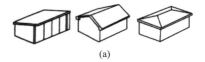

Fig. 9-5   Appearance of Roofs
(a) Pitched Roofs; (b) Curved Roofs

(5) Staircases: The transportation of people from one floor to another in a building is realized through staircases. Besides staircases, elevators and escalators are vertical transportation equipment, too. Elevators are usually used in a building with many floors or when they are needed for special use. Escalators are often used in large-scale public buildings such as

to be left inside the floor for the installing ducts of water, gas, ventilation, electric or telephone wires, etc. The materials for making floors are usually timber, reinforced concrete and steel-concrete composite as shown in Fig. 9-4. Reinforced concrete floors are used widely due to their good advantages in strength, rigidity, endurance and fireproofing.

(4) Roof: The top part of a building is termed the roof. It is not only the enclosing structures keeping the space below the roof in a good condition without the attacks of wind, rain and snow, but also the load-bearing structure supporting itself weight and various live loads. A building can be designed with different types of roofs that should be in harmony with the whole architectural style, function, structural system and covering material on the roof of the building. Flat, pitched and curved roofs are three types of roofs and Fig. 9-5 shows the latter two types. Every roof of a building needs to be constructed with a certain slope to discharge the rainwater on the roof. So the flat roof is referred to as the roof with a slope less than or equal to 10%.

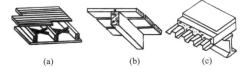

Fig. 9-4   Types of Floor Boards
(a) Timber Floor; (b) Reinforced Concrete Floor;
(c) Composite Steel-Concrete Floor

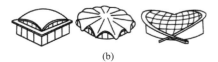

train stations, airport terminals, shopping centers, etc., where a large number of people need to go up and down. Staircases are the most basic and economical daily transportation equipment. They are also used for evacuating people in emergency structures. Staircases are therefore still necessary for buildings where elevators or escalators have been installed. Timber, brick, reinforced concrete and steel are some common materials for making staircases.

（6）Windows & doors: Both windows and doors are two kinds of enclosing structures in a building. The function of windows is to allow light and air to get in and people to see out. Doors used for people to get in and out between two separated rooms. At the same time, doors have the function of daylighting and ventilation like windows. Under different service conditions, windows and doors should serve the function of heat or sound insulation and water, fire, dust or burglary prevention. Both windows and doors can be made usually using wood, aluminum, plastic or steel. They can be opened in many ways shown in Fig. 9-6 and Fig. 9-7.

on the upstairs outside the wall of a building, with a wall or tail around it. People can get out onto a balcony from an upstairs room and enjoy the outside view or sit basking in the warm sunshine. A balcony may be arranged as shown in Fig. 9-8, where the two balconies are called cantilevered balcony and recessed veranda respectively.

Fig. 9-7   Ways of Opening Doors

Fig. 9-6   Ways of Opening Windows

（7）Balcony: A balcony is a platform built

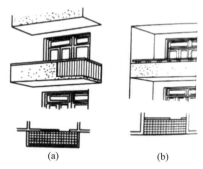

| (a) | (b) |

Fig. 9-8   Types of Balconies
(a) Cantilevered; (b) Recessed

9.4   Categories of Buildings

Buildings are often classified based on a certain criterion. The aim of classification is to investigate the feature and behavior of each type of building and then to establish standards and specifications for the design and construction of the building.

All buildings may be classified as production buildings and civil buildings according to their functions. The production buildings may be subclassed as industry buildings and buildings for agriculture and animal husbandry, and the civil buildings may also be subclassed as residential buildings and public buildings.

Industry buildings are referred to as factory buildings for manufacturing goods and auxiliary

buildings like storehouses, garages, mending workshops as well. To meet the needs of production process, most light factory buildings may be designed with multi-storey, whereas heavy factory buildings are usually characterized by single storey and large span. Agriculture and animal husbandry buildings mean the buildings such as storehouses of crops, livestock stalls, grain processing plants, tractor stations, greenhouses, etc. Residential buildings are about any types of buildings for families or a group of people to live, including houses, apartments and dormitories. Public buildings are places where people perform a variety of activities concerned with policy, diplomacy, economy, science, technology, culture, sports and so on. They play a significant role in a city, which represents the city image and reflects the living and cultural qualities of people. There are many types and complicated functions for public buildings. For example, there are many differences in architectural style, space and function among schools, hospitals, hotels, theaters, museums, libraries, railway stations, stadiums, etc. The planning and design of public buildings have much to do with the conditions of economy, science and technology as well as the aesthetic standard of the times.

　　Buildings can be constructed using different materials so that they are often classified as buildings with wood structures, masonry structures, reinforced concrete structures and steel structures. Buildings can be also be classified based on their number of storeys. For example, the residential buildings with a range of 1 to 3 storeys, 4 to 6 storeys, 7 to 9 storeys and 10 to 30 storeys are defined as low-rise buildings, mid-rise buildings mid-high-rise buildings and high-rise buildings respectively, whereas the public buildings with a total height greater than 24m belong to high-rise buildings. Any building higher than 100m is defined as a super high-rise building, regardless of a residential building or a public building.

Vocabulary and Expressions：

abutment [ə'bʌtmənt] n. 邻接，接界，桥台

anchor [ˌænkə] n. 锚；v. 固定，栓住

bore [bɔ:] n. 孔，枪膛；v. 钻孔

buttress ['bʌtris] n. 拱壁，扶壁，支持物

canal [kə'næl] n. 沟渠，运河，水道

civil ['sivl] adj. 土木工程的，民用的

collapse [kə'læps] vi. 倒塌，崩溃；vt. 使倒塌，使崩溃；n. 倒塌，崩溃，断裂

counteract [ˌkauntə'rækt] vt. 抵消，抵抗，抵制

dam [dæm] n. 水坝

deck [dek] n. 甲板

disastrous [di'za:stris] adj. 悲伤的，损失惨重的

embankment [im'bæŋkmənt] n. 堤防，筑堤

entail [in'teil] vt. 使必需，使承担，需要，遗传给

fissure ['fiʃə] n. 裂缝；v. （使）裂开，（使）分裂

flood [flʌd] n. 洪水，涨潮

fortification [ˌfɔ:tifi'keiʃən] n. 筑城，防御工事

geology [dʒi'ɔlədʒi] n. 地质学

gorge [gɔ:dʒ] n. 峡谷，凹槽

grout [graut] n. 水泥浆；v. 用水泥浆填塞

hydraulics [hai'drɔ:liks] n. 水力学

hydrology [hai'drɔlədʒi] n. 水文学

penstock ['penstɔk] n. 闸门，给水栓，水道，水渠

pier [piə] n. 码头，桥墩

porous ['pɔ:rəs] adj. 多孔的

potentiality [pəˌtenʃi'æliti] n. 潜在性，潜能

reinforce [ˌri:in'fɔ:s] v. 加强，加固，增援

soil [sɔil] n. 土壤，土地

span [spæn] n. 跨距

spillway ['spilwei] n. 溢洪道，泄洪道

survey [sə:'vei] vt. ['sə:vei] n. 测量，调查，勘定

suspender [sə'spendə(r)] n. 悬挂者（物），吊杆

suspension [səs'penʃən] n. 悬挂，暂停，中止

tunnel ['tʌnl] n. 隧道，地下道

valley ['væli] n. 山谷，溪谷，凹地

wedge [wedʒ] n. 楔子，楔形物

parapet ['pærəpit] n. 女儿墙，栏杆，扶手

canopy ['kænəpi] n. 雨篷，天篷

topography [tə'pɔgrəfi] n. 地形学

brachy ['bræki]-axis 短轴

bearing wall 承重墙；non-bearing wall 非承重墙；partition wall 隔墙；filler wall 填充墙；glass curtain wall 玻璃幕墙

flat roof 平屋顶；pitched roof 坡屋顶；curved roof 弧形屋顶

evacuate [i'vækjueit] vt., vi. 疏散，撤离

elevator ['eliveitə(r)] n. 电梯

escalator ['eskəleitə(r)] n. 自动扶梯

cantilevered balcony 悬挑阳台，凸阳台；recessed veranda 凹阳台

agriculture and animal husbandry building 农牧业建筑

truss［trʌs］n. 桁架

funicular［fjuː'nikjələ(r)］n. 缆索铁路，缆车，缆车铁道；adj. 用缆索牵引的，绳索的

folded plate 折板

low-rise buildings 低层建筑

mid-rise buildings 中层建筑

mid-high-rise buildings 中高层建筑

high-rise buildings 高层建筑

super high-rise building 超高层建筑

Notes：
① 悬索桥一般用于跨度很大的桥梁，两根锚固在两端的钢索支撑在塔上。甲板用吊杆与缆绳相连。
② 沿建筑短轴和长轴布置的墙分别称为横墙和纵墙。最外侧的横墙通常称为山墙。

Exercises：
1. Answer the Following Questions in English.
（1）What are basic elements that comprise a building? What loads may a building be subjected to?

（2）What is the function of a window and a door? What is the feature of a truss?

2. Translate the Following Paragraph into English.

The word civil derives from the Latin for citizen. In 1782, Englishman John Seaton used the term to differentiate his nonmilitary engineering work from that of the military engineers who predominated at the time. Civil engineering is the planning, design, construction and management of the built environment. This environment includes all structures built according to scientific principles, from irrigation and drainage systems to rocket launching facilities.

3. Translate the Following Paragraph into English.

土木工程领域的很多高级工程师从事设计工作。正如我们所看到的，土木工程师的工作涉及多种类型的结构，所以对一个工程师而言，仅在某种结构上有专长是一种正常现象。在建筑设计时，工程师通常是某个建筑设计公司或施工企业的顾问。堤坝、桥梁、给水系统和其他大型项目通常聘用几位工程师，在负责整个项目的系统工程师领导下协同工作。

10 Exploration of In-Class Education of Ideology and Politics of the Course "Architectural Economy and Management"

（"建筑经济与管理"课程思政探索）

教学目标: 本单元的主要内容有"建筑经济与管理"课程简介、课程思政元素、国家意识、民族认同感和自豪感、港珠澳大桥、社会责任、敬业精神、环保意识、法律意识、德育获得感以及课程思政效果。教学重点是课程思政元素，难点是课程思政效果。通过本单元的学习，要求理解国家意识、社会责任、敬业精神、环保意识、法律意识等课程思政元素，了解如何使大学生有德育获得感，以取得良好的课程思政效果。

To fully understand and apply Xi Jinping Thought on Socialism with Chinese Characteristics for a New Era, to carry out the spirit of ideological and political work conference of colleges and universities throughout China, to give full play to the role of the main channel classroom teaching in the ideological and political work of colleges and universities, and to actively promote the teaching system of "in-class education of ideology and politics", *in-class education of ideology and politics*[1] in the teaching process of the elective specialized course *"Architectural Economics and Management"*[2] in architecture major is discussed in this paper. The in-class education of ideology and politics aims to offer college students with confidence of China's building industry, inspiring national identity and sense of pride, and aspiring to become excellent architectural professionals with national consciousness, law-consciousness and environmental awareness and daring to take responsibility and have professional dedication while they acquire basic knowledge and theory of architectural engineering economy and enterprise management.

10.1   Course Introduction

The main content of this course includes three aspects: the management, economy and cost of construction projects. Students are required to master engineering construction procedure, management organization, target control and contract management, to grasp the time value computation of construction funds, to carry out the economic investment effect evaluation and uncertainty analysis of construction engineering projects, to understand composition, valuation basis and reasonable determination of the construction project cost, and to be qualified for the actual work of construction enterprise management.

According to the requirements of in-class ideology and politics, the course syllabus and the examination syllabus of "Building Economics and Management" have been revised, the allocation of lecture hours has been readjusted, the teaching plan has been modified accordingly, and a new course examination mission statement and grading standards have been completed according to the requirements of "adding ideological and political content of a course into the scope of grading assessment". Through course learning combined with professional cases, college students could further understand the core values of socialism, re-recognize their ideology, improve their ideological and moral awareness, and enlarge their sense of gain on moral education.

10.2   Ideological and Political Elements

10.2.1   National Consciousness

When delivering section "Outline of Building Industry", through the interpretation of top-ranked *Hong Kong-Zhuhai-Macao Bridge*[3] among the world's ten longest cross-sea bridges, the great achievements of Reform and Opening-up are integrated with in-class education of ideology and politics, college students' national conscious-

10   Exploration of In-Class Education of Ideology and Politics of the Course "Architectural Economy and Management"
("建筑经济与管理"课程思政探索)

061

ness and national sense of pride are improved. They are finally fiercely patriotic and still further love the Communist Party. Their confidence of China's building industry is established. Let them believe that the 21st century will be a Chinese century!

China's civil and bridge engineering technology has reached the world's top level. The height of super high-rise buildings symbolizes the height of a country and stands for the level of productivity. China has 6 of the world's 10 tallest buildings in 2018. China owns 8 of the world's 10 tallest bridges. China accounts for half of the world's 10 longest cross-sea bridges, and the Hong Kong-Zhuhai-Macao Bridge with a length of 55km ranks first in the world (Fig. 10-1).

Fig. 10-1   View of Hong Kong-Zhuhai-Macao Bridge

The Hong Kong-Zhuhai-Macao Bridge, a super-large cross-sea bridge-tunnel engineering linking Hong Kong, Zhuhai and Macao in China, opened on October 24, 2018. It takes only half an hour to drive from Hong Kong to Zhuhai or Macao. The bridge deck is a two-way six-lane expressway with a designed speed of 100 km/h, with a total investment of 126.9 billion Yuan. With a service life of 120 years, it can withstand the 8-magnitude earthquake, the 16-magnitude typhoon, the impact force of 300,000 t, and the once-in-300-years flood in the *Pearl River estuary*[④]. *Drivers drive on the bridge in the right way, and automatically adjust to the left direction after arriving at the Hong Kong or Macao port areas. Vehicles are set in the automatic conversion and interchange mode of "separate and parallel first, then overlap up and down and finally separate and parallel again" "right up and left down" or "left up and right down".*[⑤]

Hong Kong-Zhuhai-Macao Bridge is the interconnectivity "backbone" and large cross-border channel of *Guangdong-Hong Kong-Macao Large Bay Area*[⑥]. It greatly shortens the space-time distance among Guangdong, Hong Kong and Macao, thereby increases the efficient flow and configuration of innovation elements such as people, logistics, capital, technology and the like, promotes the construction of the Guangdong-Hongkong-Macao Large Bay Area becoming a more dynamic economic zone, a high-quality life circle suitable for living, working and traveling, and a demonstration zone of deep cooperation among mainland, Hong Kong and Macao, and finally builds an international high-level bay area and a world-class city group. As a milestone of China from a big to a powerful bridge nation, the bridge is known as the "Mount Everest" in bridge industry. British media, the Guardian, called it "one of the seven wonders of the modern world". It's the symbol of China's comprehensive national strength and a major measure of China's central government to support rapid development of Hong Kong, Macao and the Pearl River Delta cities. It's the important achievement of close cooperation among Guangdong, Hong Kong and Macao under "one country, two systems".

Central general secretary of the Communist Party of China, national chairman and central military commission chairman Xi Jinping assessed the Hong Kong-Zhuhai-Macao Bridge: It is an important national engineering, creates many world records, reflects the fighting spirit of building roads or bridges facing to mountains or water, embodies comprehensive national strength and independent innovation ability, and reflects national aspirations of bravely building world first-class engineering. This is a bridge of realizing Chinese dreams, uniting, self-confidence and renaissance. The completion of the bridge and its opening to traffic have further strengthened our *confidence in the path*, *theory*, *system and culture of socialism with Chinese characteristics*.[7] This fully demonstrates that both socialism and the new era rely on hard work.

10. 2. 2　Social Responsibility

When delivering residential buildings in section "Definition of Building Industry", starting from the change of Chinese housing in the past 40 years, citizens' social responsibility is explained. All citizens will have their own houses and realize the dream "home ownership". We will cultivate college students to become a new generation who are responsible for national renaissance and realizing the Chinese dream. Since Reform and Opening up, the "home" of ordinary Chinese people has undergone a major transformation from welfare housing allocation to housing marketization. Residential housing changes from the necessity of sharing a room for several generations to the "luxury" of enjoying life today. According to the data from the national bureau of statistics in July 2017, the per capita housing area in China has increased from 3.6 to 40.8 m^2.

Before Reform and Opening up, that three generations lived in one room under the welfare housing system was a true portrayal of the housing conditions at that time. *Tube-shaped apartments*[8] were the most representative housing style in the early stage of Reform and Opening up. There were countless Chinese people getting married, having children, and playing a symphony of pots, and pans in them (Fig. 10-2). In the early 1980s, "tube-shaped apartments" were gradually replaced by multi-storey residential buildings. At the beginning of 1990s, the problem of residents' housing was mainly solved by "building houses by the nation, allocating housing by organizations, and asking for housing from units". It was not easy to allocate housing. Housing reform began in the mid of 1990s, with the originally allocated "welfare housing" sold at a discount to users. This kind of sold "welfare housing" was called a housing-reform house. 1998 was a substantial year when China's housing reform took the road of complete commercialization from the era of "welfare housing system". Subsequently, residential housing area gradually increased and houses were usually multi-storey with no elevator. In the 21st century, various types of houses, such as high-grade residential areas, duplex buildings, villas, apartments and low-rent houses, have mushroomed, and the decoration level is getting higher and higher (Fig. 10-3).

Fig. 10-2　Washing Room in Tube-shaped Apartments

10. 2. 3　Professional Dedication

When delivering section "Building Industry Ethics Code", through the explanation of "Building Industry Employees' Professional Ethics Code (Trial Version)", it emphasizes that "credit and responsibility" are the most important content of the code. As future employees of building industry, college students must abide by laws and regulations, love their jobs and dedicate their lives to the industry, be incorruptible

10 Exploration of In-Class Education of Ideology and Politics of the Course "Architectural Economy and Management"
("建筑经济与管理"课程思政探索)

063

in carrying out their official duties, be honest and trustworthy, not participate in activities that interfere with official duties, and not do anything harmful to the image of the government.

Fig. 10-3   Modern Luxury Housing

10.2.4   Environmental Awareness

When delivering section "Content of Feasibility Study", that Environmental Impact Assessment (EIA) report which must be provided at the feasibility study stage of a project is firstly explained. One-vote veto system is adopted, that is, if EIA is not approved, the feasibility study report will certainly not be approved. EIA reports must be prepared by the unit qualified for EIA and be approved by environmental protection administration departments. *Then, on the basis of adhering to the construction procedure of "survey first, design later, and construction finally" "three synchronizations" must be achieved in the implementation of construction projects. These "three synchronizations" are synchronous design, synchronous construction and synchronous operation of the main project and its corresponding environmental protection measures' project.* [9] Finally, by discussing the case of "cancer-prone village", the college students' consciousness of taking good care of and protecting environment is cultivated.

"Cancer-prone village" are several examples among about 460 "cancer-prone village" in mainland China. Shaoxing County of Zhejiang Province, known as "the city built on cloth", has more than 9000 textile enterprises and its printing and dyeing capacity accounts for about

30% of the country's total. However this big hero on GDP achievement book becomes the arch-criminal of changing from beautiful waterland to heavily polluted area and from "small bridges, flowing water and homes" to "cancer-prone village". Shaoxing Binhai Industrial Zone and its periphery already have several "cancer-prone village". Sanjiang Village and Xiner Village in Shaoxing are "cancer-prone village" with high popularity in Shaoxing area. There are many printing and dyeing enterprises around, and all the inland rivers have been polluted by industrial sewage. River water has become toxic water in the eyes of villagers. "You will get a rash as long as your hands touch the water (Fig. 10-4)." Wuli Village of Xiaoshan District in Hangzhou changed from a small village with beautiful mountains and clear waters to a "cancer-prone village" in less than 20 years. Girls hate to marry with the local male villagers. "Cancer villages" seem to appear wherever chemical enterprises go.

Fig. 10-4   A Polluted River in Zhejiang

We should firmly remember Xi Jinping's green development concept. "Clear water and green mountains are just the mountains of gold and silver." "To protect ecological environment is to protect productivity. To improve ecological environment is to develop productivity. A good ecological environment is the fairest public products and the most beneficial well-being of the people's livelihood." "Only the strictest system and the strictest rule of laws can provide a reliable guarantee for the construction of ecological civilization." Under the guidance of green development concept, the construction of "Beautiful Countryside" is being carried out.

"Clear water and green mountains" in China are being changed into the mountains of gold and silver to benefit people's livelihood. The beautiful China that "can see mountains and water and remember homesickness" is approaching everyone's eyes from a scroll of landscape painting (Fig. 10-5).

10. 2. 5　Law-Consciousness

When delivering chapter "Contract Management of Building Engineering", it emphasizes that market economy is contract economy. Through explaining five typical contract breach cases of construction projects actually judged by domestic courts, the contract awareness and legal awareness of college students should be strengthened. The content of a case includes basic case information, basic details of a case, case focus, gist of court's judgment and judge's reflections. Especially case focus and judge's reflections reflect the essence of contract cases and the warnings given by judges to construction professionals.

Fig. 10-5   A Scene of Beautiful Countryside

10. 3　Ideology and Politics Education Effect

Through the introduction of Hong Kong-Zhuhai-Macao Bridge, the national consciousness and aspirations of college students have been enhanced so as to make them realize that everyone is responsible to make China becoming a powerful country. We should not only be highly confident and good at learning, but also have the courage to innovate. Happiness comes from struggle and a powerful country with modern civilization comes from hard work. By means of explaining residential building change in 40 years of Reform and Opening up, college students realize that they should undertake social responsibility, have courage to take responsibility, and struggle for the realization of the dream "home ownership".

By expounding "Building Industry Employees' Professional Ethics Code (Trial Version)", they realize that building industry employees' professional ethics are the concrete embodiment of socialist core values in building industry. Their professional ethics, professional dedication and personal cultivation in building industry in the future are further improved.

While narrating EIA, the painful lessons of "cancer villages" have improved their environmental awareness, and have strengthened the concepts of green GDP, one-vote veto system of EIA, the construction procedure of "survey first, design later and construction finally" and "three synchronizations" of construction projects.

Several typical contract breach cases of construction projects actually judged by courts make them realize that when signing and executing contracts of engineering projects, the qualifications of design institutes and construction contract enterprises must be strictly reviewed, attention should be paid to the rigor and accuracy of contract wording, advance and progress payments must be paid strictly in accordance with the contract agreement, the delay reasons of work period must be timely found out and confirmed, and stress must be laid on the evidence procurement and preservation of work period delay, workload change and the like lest they should go to court in the future.

10 Exploration of In-Class Education of Ideology and Politics of the Course "Architectural Economy and Management"
（"建筑经济与管理"课程思政探索）

065

Vocabulary and Expressions：

ideology [ˌaidiˈɔlədʒi] n. 意识形态，思想体系，思想意识

Xi Jinping Thought on Socialism with Chinese Characteristics for a New Era 习近平新时代中国特色社会主义思想

national identity and sense of pride 民族认同感和自豪感

national consciousness 国家意识

environmental awareness 环保意识

law-consciousness 法律意识

engineering construction procedure 工程项目建设程序

time value of construction funds 建设资金时间价值

economic investment effect evaluation 投资经济效果评价

uncertainty analysis 不确定性分析

construction project cost 建筑工程造价

course syllabus 课程教学大纲

examination syllabus 课程考试大纲

core values of socialism 社会主义核心价值观

sense of gain on moral education 德育获得感

professional dedication 敬业奉献

super high-rise building 超高层建筑

logistics [ləˈdʒistiks] n. 物流

one country, two systemsg 一国两制

the Chinese Dream 中国梦

welfare housing 福利房

a housing-reform house 房改房

duplex building 复式楼

low-rent house 廉租房

Environmental Impact Assessment（EIA）report 环境影响评价报告（环评报告）

feasibility study 可行性研究

work period delay 工期延误

workload change 工程量变更

Notes：

① 课程思政。

② 本科建筑学专业的一门专业选修课程"建筑经济与管理"。

③ 港珠澳大桥。

④ 珠江口。

⑤ 桥上驾驶采用右行规则，到达港澳口岸地区后顺道路方向自动调整为左行规则，车辆按照"先分离并行，再上下重叠，最后又分离并行"和"右上左落"或"左上右落"立交自动转换方式进行设置。

⑥ 粤港澳大湾区。

⑦ 中国特色社会主义的道路自信、理论自信、制度自信和文化自信。

⑧ 筒子楼。

⑨ 在建筑工程项目实施过程中坚决遵循"先勘察、后设计、再施工"的建设程序，严格做到"三同步"，即主体工程与环保措施工程同步设计、同步施工、同步投入运行。

Exercises：

1. Translate the Following Paragraph into Chinese.

The contractor's bid estimates often reflect the desire of the contractor to secure the job as well as the estimating tools at its disposal. Some contractors have well established cost estimating procedures while others do not. Since only the lowest bidder will be the winner of the contract in most bidding contests, any effort devoted to cost estimating is a loss to the contractor who is not a successful bidder. Consequently, the contractor may put in the least amount of possible effort for making a cost estimate if it believes that its chance of success is not high.

2. Translate the Following Paragraph into English.

如果一个总承包商在项目承建过程中想要转包，对转包给次承包商的不同的任务要作出报价单。这样，总承包商将把成本估价的风险转移给次承包商。如果总承包商承担全部或部分建造工作，投标报价将根据业主所提供的图纸或根据执行合同的承包商设计的施工程序来确定。例如，在有关成本资料的商业出版物中可以找到某种类型或规模的基础费用，这些成本资料便于由工程量清单做成本估价。但是，如果认为项目不同于典型的设计，承包商要考虑采用实际的施工程序和相关的费用，来评定实际建筑成本。因而，完成各项任务所需要的，像劳动力、材料和设备之类的因素将用作成本估价的参数。

11　Civil Engineering Materials
（土木工程材料）

教学目标：本单元的主要内容有各种土木工程材料，如木材、金属、混凝土、沥青、聚合物、砖、砌块等的特性及其使用。教学重点是金属、混凝土和砖，难点是混凝土材料的特性。通过本单元的学习，要求掌握各种土木工程材料性能和优缺点。

Civil engineering materials can be natural and manmade. They contain rock, metals, timber, concrete, bituminous, soil, polymers, bricks, blocks, etc. Besides these traditional materials, new types of constructional materials are also investigated and developed and will be applied gradually. Now green civil engineering materials and even eco-materials for civil engineering are recommended based on the consideration of Sustainable Development. This has the benefits of reducing energy, saving resources and protecting the environment, having minimum harm to human health.

11. 1　Timber

11. 1. 1　Features of Timber

Timber is one of the oldest known civil engineering materials. In addition to its usefulness as a structural material, timber has also fulfilled a role in temporary structures. Although timber is a kind of sustainable resource, the consumption speed of forests must be slowed down because of the relative slowness of tree growth. Mature trees, of whatever type, are the source of structural timber. *From the cross section of trunk (Fig.* 11-1*), we can observe the bark, heartwood, pith, sapwood and growth rings.* [1] Timber is generally classified into two types: (1) hardwoods which have broad leaves shed in the autumn, such as angiosperm and dicotyledons; (2) softwoods which have needle-like leaves, broadly evergreen, such as conifers and gymnosperms. It should be noted that the termshardwoods and softwoods do not necessarily indicate relative hardness or density. The easy growth of softwoods makes softwoods commercially much cheaper than hardwoods, which have a slower rate of growth and do not generally grow in profusion in northern hemisphere countries.

The principal characteristics of timber are strength, durability and finished appearance. Strength of timber is affected by factors such as density, moisture content and grain structure as well as by various defects. Density is almost certainly an indication of strength: the denser the timber is, the stronger it is. For softwoods, the growth rate has major strength significance. For various species there are optimum growth rates and it has been shown that timber which has not matched or has exceeded these optima is likely to be weaker. A rough rule of thumb for most timbers is that a range of 6 to 13 rings per 25 mm measured radially can be considered to indicate reasonable strength. The upper value of the range generally gives greater strength in softwoods while the lower value of the range is preferable for ring-porous hardwoods.

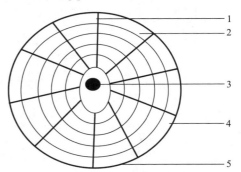

Fig. 11-1　Cross Section of the Trunk
1-Pith; 2-Growth Rings; 3-Heartwood; 4-Sapwood; 5-Bark

Moisture content of timber is the quantity of moisture contained expressed as a percentage of the dry weight. The moisture content plays a great role in affecting the strength of timber. The strength generally decreases with a rise in moisture content. There is optimum moisture content and an excessively dried-out timber may have a lower strength. Grain structure and continuity are of significance in a strength context and any disruption due to growth defects or seasoning defects will induce a reduction in strength.

11.1.2 Defects in Timber and Preservation of Timber

Defects can occur in timber at various sta-

ges of its growth, as well as during its conversion and seasoning process. Any of these defects can cause trouble in use, either by reducing its strength and durability, or marring its appearance（Tab. 11-1）.

It is necessary for timber to be preserved or treated to resist infestation and decay. There are many types of preservation methods：1）tar-oil preservatives；2）water-soluble preservatives；3）organic solvent preservatives；etc.

Timber Deterioration Sources and Forms **Tab. 11-1**

Source		Form of deterioration
Bacteria		Discoloration of surface
Chemicals		Discoloration of surface with disintegration of cellular structure
Fire		Chairing of surface of thick section
Fungi	Moulds	Discoloration of surface
	Dry rot	Discoloration of surface with disintegration of cellular structure
	Microscopic rots	
	Wet rots	
Insect infestation	Marine borers	Flight-holes at surface with tunneling causing disintegration of structure
	Beetles	
Mechanical	Loading	Fracture of fibers
	Abrasion	Surface deterioration
	Erosion	
Sunlight		Discoloration embrittlement of surface
Water	Flowing	Discoloration and erosion of surface
	Intermittent	Expansion and contraction leading to cracking and splitting of cellular structure

11.1.3 Use of Timber

Timber is widely used in permanent carcass and structures. A range of such uses is given below. （1）Marine work：Much of the marine work traditionally requires timber for wharves, piers, sheet piling, cofferdams, etc. For timber used in marine work, requirements are high density, close grain structure and a natural resistance to impact, infestation, fungal attack, salt or wave erosion and temperature variation； （2）Heavy

constructional work：Among other uses under this heading are pylons, gantries, bridges, shoring, abutments and so on. In many instances of heavy work, requirements are the same as for marine work. High density, closeness of grain and resistance to impact are all important, with the addition of resistance to acidity, alkalinity or other chemical nature of the soil with which the timber will be in contact；（3）Medium/light constructional work：Roof trusses, partitions, screens, floors and walls are factory produced partially with timber. When timber is used for these purposes, requirements on resistance to insect and fungal attack, and low dimensional change due to temperature and humidity, are very important. Because the location in this type of work will almost certainly be inside a building, it may be necessary to reinforce any natural resistance to fire or rate of flame spread by impregnation with a fire-retardant chemical.

The above relates to the use of timber in structural carpentry. Indeed, nowadays timber is also playing an important role in falsework carpentry, such as shuttering for in-situ or precast concrete work, supporting formwork for brick or stone arch or shell forms, or jigs for glued-laminated timber beam or shell forms. Timber has a wide use in flooring, facing, skirting, windows, doors, stairs, paneling and furniture. The requirements for this purpose include ease of working and finishing, good grain pattern and appearance when clear-finished, dimensional stability in conditions of variability of temperature and humidity, both internal and external, and resistance to infestation, fungal attack, etc. Small section timber is often used to build up much larger components by the use of adhesives. These components derive strength and functional properties from the enhanced permissible stresses generally associated with this laminated form of construction, the availability of suitable adhesives and the geometry of built-up forms.

11.2　Metals

11.2.1　Properties of Metals

Metals consist of ferrous metals and nonferrous metals. The ferrous metals mean iron and alloys made mainly from iron (iron alloys), while the nonferrous metals include all the other metals and their alloys, such as copper, aluminum, magnesium, aluminum alloy, aluminum bronze, etc. The applications of metals in civil engineering are wide and varied, ranging from their use as main structural materials to their use for fastenings and bearing materials. As main structural materials cast iron and wrought iron have been superseded by rolled-steel sections. Steel is also of major importance for its use in reinforced and prestressed concrete.

The properties of metals which make them unique among constructional materials are high tensile strength, the ability to be formed into plate, sections and wire and the weldability. Other properties of metals are electrical conductivity, high thermal conductivity and metallic luster, which are of importance in some circumstances. Perhaps the greatest disadvantage of the common metals and steels in particular, is the need to protect them from corrosion by moist conditions and atmosphere, although weathering steels have been developed which require no protection from atmospheric corrosion.

When in service, metals frequently have to resist not only high tensile or compressive stress and corrosion, but also conditions of shock loading and low temperatures. Steel does not meet the demands where the properties of high electrical conductivity or corrosion resistance are required. Normally one or more alloying elements are added to increase strength or to modify the properties. The importance of metals as constructional materials is almost invariably related to their loadbearing capacity in either tension or compression and their ability to withstand limited deformation without fracture. It is usual to assess these properties by tensile tests in which the modulus of elasticity, the yield or proof stress, the tensile strength and the percentage elongation can be determined.

11.2.2　Use of Metals in Constructions

Many ferrous and nonferrous metals and alloys are available to engineers. Iron alloys are used in large quantities than the alloys of any other metal. This arises from the relative cheapness with which steels and cast irons can be produced with a variety of useful properties.

(1) Wrought steels: Engineering components are produced in these steels by hot or cold working processes. The components may be put directly into service or may be heated or surface treated depending on the grade of steel or the application.

(2) Structural steels: Structural steels are main materials in steel structures. Design of steel structures is based primarily on the strength of the steel but ductility, toughness and weldability are often important properties. The weldability of steels deteriorates with the increase of carbon content. The weldability of steels is very important because welded structures give a weight saving and ease of fabrication compared with bolted or riveted structures.

(3) Plain carbon steels: Plain carbon steel, also called mild steel, has good ductility and weldability if its carbon content is low and has a high yield strength if its carbon content is high. Consequently, the carbon content is limited to 0. 38% maximum in the basic structural steels to give a compromise between the opposing requirements. The carbon steels also contain the elements manganese, silicon, Sulphur

and phosphorus that arise and are controlled in the steelmaking process. Manganese improves strength and ductility, and the content between 0. 5% and 1. 0% is normally present. Silicon improves the strength but excessive amounts may cause the carbon to occur as graphite flakes which reduce the strength, and so the silicon content rarely exceeds 0. 6%. Sulphur and phosphorus embrittle the steel and are controlled to 0. 05% maximum.

(4) Low alloy steels: Low alloy steels contain more alloy content than plain carbon steels. The alloy elements are manganese (0. 8% ~ 1. 7%), silicon (≤ 0. 50%), vanadium (0. 02% ~ 0. 20%), niobium (0. 013% ~ 0. 06%), titanium (0. 02% ~ 0. 20%), etc. Compared with the plain steels, low alloy steels have higher strength, better ductility and same weldability. Low alloy steels are usually used in tall buildings, long span building, high-size structures, long span bridges, heavy industrial factories, etc.

(5) Concrete reinforcement: As we know, concrete has low tensile and bending strength compared to its high compressive strength, and concrete is easy to crack even under a very low stress. Steel reinforcement concrete can overcome the deficiencies in the tensile and bending strengths of concrete (Fig. 11-2).

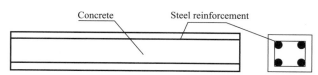

Fig. 11-2   Steel Reinforced Concrete

The reinforcing steel must have adequate tensile properties and form a strong bond with the concrete. The bond force is purely mechanical and arises from surface roughness and friction of the reinforcement. Mild steel with a maximum carbon content of 0. 25% is suitable and supplied in three conditions. These are hot rolled, cold rolled and hard drawn. Pre-stressing steels must have a high yield stress in ten-

sion so that a high tensile force can be inducted in them. High strength steel is mainly used for prestressing steels.

(6) Aluminum and aluminum alloys: The useful engineering properties of both unalloyed and alloyed aluminum are low specific gravity, resistance to corrosion, high electrical conductivity and excellent forming properties. The low strength of aluminum is a disadvantage and for

satisfactory service it must be alloyed and mechanically or thermally treated to give improved strength.

Super-purity 99.99% aluminum is too costly for general engineering applications and commercial grades, which vary in purity from 99.0% to 99.8% aluminum. It is normally selected on the requirements of corrosion resistance, formability or tensile strength. Corrosion resistance and formability are enhanced and tensile strength is reduced by increased purity. Strengthening may be achieved by cold working and softening by annealing at various temperatures to produce conditions varying from fully work hardened to fully softened. Alloy additions are chosen so that the strength can be enhanced without an adverse effect on the specific gravity or corrosion resistance. The alloying elements commonly added to aluminum are copper, manganese, magnesium, silicon, nickel and iron. Part of the addition forms a solid solution with the aluminum and part forms a compound with the aluminum or with one of the other alloying elements. The corrosion resistance of an alloy is good if the solid solution and associated compound have similar chemical properties. Alloys containing copper, nickel and iron have poorer corrosion resistance than those with magnesium, silicon and manganese.

11.2.3　Corrosion Protection

The atmospheric corrosion resistance of steel structures may be enhanced in several ways: (1) Protective coatings of paint, or metal coatings. Metal coatings have stronger resistance against corrosion than steel in all but most severe environments and possess more abrasion resistance than painted coats; (2) Cathodic protection may be applied to parts of structures which are continuously immersed in water but this is not effective when only part of the structure is immersed, for example, in a tidal situation; (3) Steels containing an increased amount of phosphorus and some chromium and copper, have a better corrosion resistance compared to normal steels. *Exposure to the weather for about two years causes the formation of an adherent protection oxide film with a pleasing purplish-copper color instead of the normal flaky rust. To ensure uniform weathering the sections are shot-blasted at works.* [2] The cost of weathering steels is about 20% higher than for normal steels but is offset by saving in weight, protective treatment and maintenance.

11.3　Concrete

11.3.1　Properties and Uses of Normal Concrete

Concrete is a man-made composite, the major constituent of which is natural aggregate (such as gravel and sand) and binding medium (such as cement paste, bitumen and polymers). The binding medium is used to bind the aggregate particles together to form a hard composite material. In its hardened state, concrete is a rock-like material with a high compressive strength. By virtue of the ease with which fresh concrete in its plastic state may be moulded into any shape, it may be used for decorative purposes.

Normal concrete has a comparatively low tensile strength and for structural applications it is normal practice either to incorporate steel bars to resist any tensile forces (steel reinforced concrete) or to apply compressive forces to the concrete to counteract these tensile forces (pre-stressed concrete or post-stressed concrete). Concrete is also used in conjunction with other materials, for example, it may form the compression flange of a box section and the remainder of which is steel (composite construction). Concrete is used structurally in buildings, shell structures, bridges, sewage-treatment works, railway sleepers, roads, cooling towers, dams, chimneys, harbors, off-shore structures, coastal protection works and so on. It is also used for a wide range of precast concrete products which include concrete blocks, cladding panels, pipes and lamp standards.

The impact strength, as well as the tensile strength, of normal concrete is low and this can be improved by the introduction of randomly orientated fibers into the concrete. Steel, polypropylene, asbestos glass, carbon and even wood fibers have all been used with some success in precast prod-

ucts and in-situ concretes, including pipes, building panels and piles. Concrete requires little maintenance and has good fire resistance. Concrete has other properties which may on occasions be considered less desirable, for example, the time-dependent deformations associated with drying shrinkage and other related phenomena.

11.3.2 Preparation of Normal Concrete

We can learn how a structural concrete is prepared from Fig. 11-3. Curing after the placing of concrete is very important for obtaining good properties.

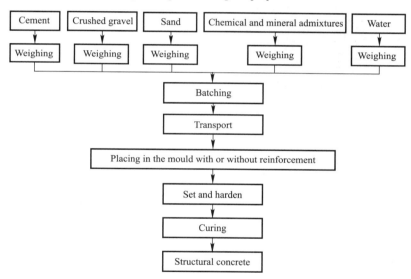

Fig. 11-3 Concrete: from Raw Materials to Structures

Making good concrete is difficult without addition plasticizers or superplasticizers.

Good concrete has to be satisfactory in its hardened state (i.e., needed strength and good durability), and also in its fresh state while being transported from mixers and placed in formworks (sometimes through pumping).

11.3.3 Constituents in Concrete

Concrete is composed mainly of three materials, namely, cement, water and aggregate

and an additional material, known as a chemical admixture, is sometimes added to modify certain of its properties. Fly ash, ground blast-furnace slag powder and silica fume are often used to replace partial cement to modify the properties of concrete and reduce the cost of concrete.

Cement is a chemically active constituent but its reactivity is only brought into effect with mixing water. The aggregate plays no part in chemical reactions but its usefulness arises because it is an economical filler material with good resistance to volume changes which take place within the concrete after mixing, and it improves the durability of concrete. A typical structure of hardened concrete and the proportions of the constituent materials encountered in most concrete mixes are shown in Fig. 11-4. In a properly proportioned and compacted concrete the voids are usually less than 2%. The properties of concrete in its fresh and hardened state can show large variations depending on the type, quality and proportions of the constituents.

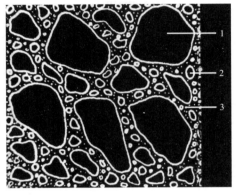

Fig. 11-4　Section Diagram of Concrete
1-Coarse Aggregate; 2-Fine Aggregate; 3-Cement Paste

(1) Cement: The different cements for making concrete are finely ground powders and all have the important property that when mixed with water a chemical reaction (hydration) takes place. This in time produces a very hard and strong binding medium for aggregate particles. In the early stages of hydration, cement mortar gives fresh concrete its cohesive properties.

(2) Aggregate: Aggregate in concrete can be classified as coarse aggregate and fine aggregate according to its diameter. The diameter of the coarse and fine aggregate is larger than 5mm and smaller than 5mm respectively. The coarse aggregate used in concrete is often cobblestone and crushed gravel and the fine aggregate is river sand and mountain sand. Aggregate is much cheaper than cement and maximum economy is obtained by using as much aggregate as possible in concrete. Its use also considerably improves both the volume stability and the durability of the resulting concrete. The physical characteristics and in some cases the chemical composition of aggregate affect to a varying degree of the

properties of concrete in both plastic and hardened state.

Sand is generally considered to have a lower size limit of about 0.08mm and a little less. Materials which size is smaller than 0.08mm are classified as clay which does harm to the mechanic properties and durability of concrete. Sea sand is not permitted to be used for making structural concrete before decreasing its salt content to less than 0.6% by washing.

(3) Water: In general water fit for drinking, such as tap water, is acceptable for mixing concrete. The impurities that are likely to have an adverse effect when present in appreciable quantities include silt, clay, acids, alkali, other salts and organic matter and sewage. The use of seawater does not appear to have any adverse effect on the strength and durability of Portland cement concrete but it may cause surface dampness, efflorescence and staining and should be avoided where concrete with good appearance is required. Seawater also increases the risk of corrosion of steel and its use in reinforced concrete is not permitted.

(4) Chemical admixtures: Chemical admixtures are chemical substances introduced into a batch of concrete, during or immediately before its mixing, in order to alter or improve the properties of the fresh or hardened concrete or both. Nowadays, there are more than 13 types of chemical admixtures produced and they benefit concrete structures in many ways. Since chemical admixtures may also have detrimental effects, their suitability for a particular concrete should be carefully evaluated before use, based on the knowledge of their main active ingredients, on available performance data and on trial mixes.

(5) Mineral admixtures: Fly ash, slag powder and silica fume can be used to replace a part of cement to reduce concrete cost and enhance concrete properties, especially its durability. Nowadays, chemical admixture and mineral admixture have become the two necessary raw materials for preparation of high-performance concrete, the fluidity and durability of which is both much better than that of normal concrete.

11. 3. 4　Concrete Mix Design

While designing a concrete mix proportion, the following factors must be carefully considered: (1) Workability of the fresh concrete; (2)Strength of the hardened concrete; (3)Durability of the concrete; (4) Cost of the concrete. Because of its complication, it should be noted that the concrete mix proportion designed should be tested and modified before being used in practice.

11. 4　Reinforced Concrete

Reinforced concrete has steel bars that are placed in a concrete member to carry tensile forces. These reinforcing bars have wrinkles on the surfaces to ensure a bond with the concrete. Although reinforced concrete was developed in many countries, its discovery usually is attributed to Joseph Monnier, a French gardener, who used a wire network to reinforce concrete tubes in 1868. This process is workable because steel and concrete expand and contract equally when the temperature changes. If this is not the case, the bond between the steel and concrete will be broken by a change in temperature since the two materials would respond differently. Reinforced concrete can be molded into innumerable shapes, such as beams, columns, slabs and arches, and is therefore easily adapted to a particular form of building. The detailed contents of reinforced concrete are shown in Unit 14 of this book.

11. 5　Prestressed Concrete

Concrete is strong in compression, but weak in tension: Its tensile strength varies from 8% to 14% of its compressive strength. Due to such a low tensile capacity, flexural cracks develop at early stages of loading. In order to reduce or prevent such cracks from developing, a concentric or eccentric force is imposed in the longitudinal direction of the structural element. This force prevents the cracks from developing by eliminating or considerably reducing the tensile stresses at the critical midspan and support sections at service load, thereby raising the bending, shear and torsional capacities of the sections. The sections are then able to behave elastically, and almost the full capacity of the concrete in compression can be efficiently utilized across the entire depth of the concrete sections when all loads act on the structure.

Such an imposed longitudinal force is called a prestressing force, i.e., a compressive force that prestresses the sections along the span of the structural element prior to the application of the transverse gravity dead and live loads or transient horizontal live loads. The type of prestressing force involved, together with its magnitude, is determined mainly on the basis of the type of system to be constructed and the span length and slenderness desired. Since the prestressing force is applied longitudinally along or parallel to the axis of the member, the prestressing principle involved is commonly known as linear prestressing.

Prestressed concrete is not a new concept, dating back to 1872, when P. H. Jackson, an engineer from California, patented a prestressing system that used a tie rod to construct beams or arches from individual blocks. After a long lapse of time during which little progress was made because of the unavailability of high-strength steel to overcome prestress losses,

R. E. Dill of Alexandria, Nebraska, recognized the effect of the shrinkage and creep (transverse material flow) of concrete on the loss of prestress. He subsequently developed the idea that successive post-tensioning of unbonded rods would compensate for the time-dependent loss of stress in the rods due to the decrease in the length of the member because of creep and shrinkage. In the early 1920s, W. H. Hewett of Minneapolis developed the principles of circular prestressing. He hoop-stressed horizontal reinforcement around walls of concrete tanks through the use of turnbuckles to prevent cracking due to internal liquid pressure, thereby achieving watertightness. Thereafter, prestressing of tanks and pipes developed at an accelerated pace in the United States, with thousands of tanks for water, liquid and gas storage built and much mileage of prestressed pressure pipe laid in the two to three decades that followed. The detailed contents of prestressed concrete are shown in Unit 14 of this book.

11.6   Bituminous Materials

11.6.1   Kinds and Properties of Bituminous Materials

Engineers have made use of the excellent durability and adhesive properties of bituminous materials. The bituminous materials used in ancient times are natural occurring bitumens. However, the bulk of today's bituminous materials are the products of industrial refining process. In fact, so-called bituminous materials include bitumen and tar. Bitumen is obtained by the fractional distillation of petroleum (crude oil). It is a viscous liquid or a solid, consisting essentially of hydrocarbons and their derivatives, which is soluble in carbon disulphide, is substantially nonvolatile and softens gradually when heated. It is black or brown in color and possesses waterproofing and adhesive properties.

Tar is a viscous liquid, black in color, with adhesive properties, obtained from the destructive distillation of coal, wood, shale, etc. No specific source is stated and it is implied that the tar is obtained from coal. All bituminous materials are for the most part in mixtures with mineral or other aggregate. Bituminous materials have the following properties and advantages: (1) relative cheapness and availability in large quantities; (2) durability; (3) good adhesive and waterproofing qualities; (4) the ease with which they can be handled at elevated temperatures, but quickly become stiff and resistant to deformation at normal temperatures.

11.6.2   Uses of Bituminous Materials

The earliest known uses of bitumen and tar relate to hydraulic uses, for example, bitumen is used to waterproof a building floor. Thin coating of bitumen paints or emulsions applied to absorptive materials have the effect of sealing capillaries so that both water and water vapor are prevented from moving through the materials. Bitumen and tar is a good material of chemical-attack resistance. It can also be painted on to parts which are liable to corrosion by weather or other sources.

Nowadays the main use of bitumen is in road surfaces, named bitumen concrete road. In order to improve the strength of bituminous materials at the high temperatures and the toughness of them at temperatures below zero, polymer modified bitumens, such as *SBS rubber modified bitumen*[3] and *APP plastics bitumen*[4] are widely used. As with most oily substances, bitumens and tars will burn if they are hot enough and there is air present. There are two temperatures which are very important for engineers to avoid to safely use bituminous materials. These are called the "flash point" and the "fire point".

11.7　Polymer

A polymer is a large molecule containing hundreds or thousands of atoms formed by combining one, two or occasionally more kinds of small molecule (monomers) into a chain or network structures. The polymer materials are a group of carbon-containing (organic) materials which have macromolecular structures of this sort. The main polymer materials which are widely used in civil engineering are PE (polyethylene), PP (polypropylene), IR (polyisoprene rubber), PVC (polyvinyl chloride), CPVC (chlorinated polyvinyl chloride) and ABS (acrylonitrile-butadiene-styrene copolymer), etc. In the last 30 years, polymers have secured a place alongside metals and ceramics as one of the major classes of manufactured materials. There are now a great variety of such materials based on about fifty individual synthetic polymers, and they find wide application throughout manufacturing industry and engineering. These materials have a spread of engineering properties which are very different from those found in metals and ceramics and a distinct manufacturing and processing technology centered largely on moulding, extrusion and fiber-forming operations.

Building construction is one of the main volume markets for polymers. However, the use of polymers in civil engineering is less conspicuous, because these materials do not generally compete directly with the traditional load-bearing materials, the structural metals, concrete and masonry. But a number of polymers have important and established civil engineering uses. One of the most straightforward and prominent applications is in pipework (Fig. 11-5), where longer established materials like cast iron, heavy clay and concrete face severe competition with several polymer materials. Polymers also play important roles as surface coatings, membranes, adhesives and jointing compounds, roofing materials, claddings and thermal insulators. Fiber-reinforced plastics play a great role in light structures. PVC doors and PVC windows have been widely used in the buildings (Fig. 11-6). They are light in weight and their thermal insulation and air impermeability are much better than wooden or metal doors/windows.

Fig. 11-5　Pipes Made of Polymers

Fig. 11-6　PVC Doors and Windows

11.8　Bricks and Blocks

Clay, calcium silicate and concrete products all include both bricks and blocks. The distinction between the two is primarily the size, i.e., blocks being larger than bricks. Clay and calcium silicate bricks and concrete blocks are widely used. All bricks and blocks have broadly similar uses although their properties differ in some important respects depending on the raw materials used and the method of manufacture. The use of common clay bricks is no longer permitted in large cities in China today because the manufacture of them will destroy arable land and seriously pollute air. The main recommended bricks and blocks in buildings are perforated clay bricks, hollow bricks, autoclaved lime-sand bricks, calcium silicate bricks, gas-entrained concrete blocks and foamed concrete blocks, etc.

Vocabulary and Expressions：

hardwood ['haːdwud] n. 硬木
angiosperm ['ændʒiə(u)spəːm] n. [植] 被子植物
dicotyledon [ˌdaikɔti'liːdən] n. [植] 双子叶植物
softwood ['sɔftwud] n. 软木
conifer ['kɔnifə(r)] n. [植] 针叶树
gymnosperm ['dʒimnə(u)spəːm] n. [植] 裸子植物
mar [maː] vt. 破坏，损毁，损伤，糟蹋，玷污，损坏…外表；n. 污点，瑕疵
fungi ['fʌŋgi] n. 真菌
carcass ['kaːkəs] n. 骨架，尸体，残骸
pylon ['pailən] n. 桥塔，指示塔，高压线铁塔
gantry ['gæntri] n. 桶架，构台，起重机架
ferrous metal 黑色金属
nonferrous metal 有色金属
vanadium [və'neidiəm] n. [化学] 钒
niobium [nai'əubiəm] n. 铌
formability [fɔːmə'biliti] n. 成形性，成型性能；[材] 可成形性
polypropylene [ˌpɔli'prəupəliːn] n. [高分子] 聚丙烯
plasticizer ['plæstisaizə] n. 塑化剂
efflorescence [ˌeflə'resns] n. 盐析，泛碱，渗斑
carbon disulphide 二硫化碳

IR（polyisoprene rubber）聚异戊二烯橡胶，异戊橡胶

Notes：
① 从树干的横截面（图11-1）可以看到树皮、心材、髓、边材和年轮。
② 暴露在天气中大约两年，会形成一层附着的保护氧化膜，具有令人愉悦的紫铜色，而不是正常的片状锈。为了保证均匀的风化，在施工中对各构件进行了喷丸处理。
③ SBS（Styrene ['stairiːn]-Butadiene [ˌbjuːtə'daiiːn]-Styrene）苯乙烯-丁二烯-苯乙烯，SBS 橡胶改性沥青。
④ APP（Abnormal PolyPropylene [ˌpɔli'prəupəliːn]）无规聚丙烯，APP 塑料沥青。

Exercises：
1. Answer the Following Questions in English.
(1) How does the moisture content in timber affect the strength of timber?
(2) What measures can we take to protect metal from corrosions?
(3) What are the main constituents in concrete? Discuss the main factors that influence the concrete properties.

2. Translate the Following Sentences into Chinese.
(1) New alloys have further increased the strength of steel and eliminated some of its problems, such as fatigue, which is a tendency for it to weaken as a result of continual changes in stress.
(2) Concrete is very versatile; it can be poured, pumped, or even sprayed into all kind of shapes.
(3) Concrete and steel also form such a strong bond — the force that unites them — that the steel cannot slip within the concrete.
(4) The lightweight aggregate reduces dead load of the floor by 20% resulting in considerable savings in the floor steel in every floor and the roof, as well as in the column steel and (less) in the foundations.

3. Translate the Following Sentences into English.

（1）新规范将边长为 150mm 混凝土立方体在 28d 龄期的抗压强度规定为混凝土的抗压强度等级，单位是"N/mm^2"。

（2）混凝土试件的抗拉试验并不容易，因为在很小荷载作用下试件就破坏，试验过程的误差会对试验结果产生显著的影响。

（3）混凝土的徐变与混凝土的组成成分、构件的尺寸、环境条件、应力水平、混凝土的龄期以及荷载持续时间有关。

12　Soil Mechanics and Foundations
（土力学与基础）

教学目标：本单元的主要内容有土力学定义、土的颗粒组成、土的结构、土的性质、土力学应用、基础分类、浅基础、深基础、桩基、摩擦桩、端承桩、钻孔桩以及沉箱。教学重点是土力学定义和基础分类，难点是沉箱基础。通过本单元的学习，要求熟悉土力学定义、土的颗粒组成、土的结构、土的性质和土力学应用，掌握摩擦桩和端承桩的定义以及基础分类，了解桩基、钻孔桩以及沉箱基础的性能。

According to Terzaghi (1948)："*Soil Mechanics is the application of laws of mechanics and hydraulics to engineering problems dealing with sediments and other unconsolidated accumulations of solid particles produced by the mechanical and chemical disintegration of rocks regardless of whether or not they contain an admixture of organic constituent.*"[1] Foundation engineering is the branch of engineering which is associated with the design, construction, maintenance and renovation of footings, foundation walls, pile foundations, caissons and all other structural members which form the foundations of buildings and other engineering structures. It also includes all engineering considerations of underlying soil and rock as they are associated with the foundation. In many countries, limited availability of space, especially in urbanized areas, has led to the need to reclaim land from the sea or build on ground with marginal or poor foundation conditions. Until recently in Shanghai, most major structures to be constructed on loose, compressible soils are founded on piles or other types of deep foundations.

12. 1　Size Composition and Structure of Soils

12. 1. 1　Size Composition

Soils are normally composed of solid, liquid and gaseous phases. Grain size is the basis of soil mechanics, since it decides whether a soil is frictional or cohesive, i.e., sand or clay. *Starting with the largest sizes, boulders are larger than 10cm, cobbles are from 5 to 10cm, gravel or ballast is from about 5 to 50mm, grit is from about 2 to 5mm and sand is from 0. 06 to 2mm. All these soils are frictional, being coarse and thus non-cohesive. Their stability depends on their internal friction. For the cohesive or non-frictional soils, the two main internationally accepted size limits are: silt from 0. 002 to 0. 02mm and clay for all finer material.*[2] There are, of course, many silty clays and clayey silts. A soil may be composed of only one size fractions in continuous or gap grading. The size composition of a soil is called its texture. A naturally occurring soil sample may have particles of various sizes.

For the purpose of defining the physical and index properties of soils, it is more convenient to represent the soil skeleton by a block diagram or phase diagram as shown in Fig. 12-1. Soils are three-dimensional systems, having a two-dimensional area extent and a third dimension, depth. Whether they are geologic deposits or formed on site by the interaction of geologic parent material, climatic factors, topography and living organisms, soils show area limitations and changes with depth.

12. 1. 2　Structure

Soil structure refers to the mutual interaction of soil particles and their arrangement in space. Non-coherent soils, such as sands and gravels, form single grain structures. Coherent soils, such as silt and clay, may form massive structures if they suffer no volume change with change in moisture content or if they are so located that no such changes occur except for continued consolidation of water-saturated sediments. Soil structure from macro to ultra-fine influences or even dominates many engineering properties such as permeability, bearing capacity, shear resistance, workability, thermal and electrical conductivity.

Generally, in undisturbed soil, four types of structures are visible. Granular structures predominate on the top, followed in succession by platy, blocky and finally prismatic struc-

tures. These layers make up what is called the soil profile. *It takes a long time to form topsoil structure, so we need to protect and conserve soil. A rule of thumb is that it takes a hundred years to form one inch (1 in = 25. 4mm) of topsoil structure. A hundred years! Remember, just as we need good soil today, so will people in the years to come.* [3]

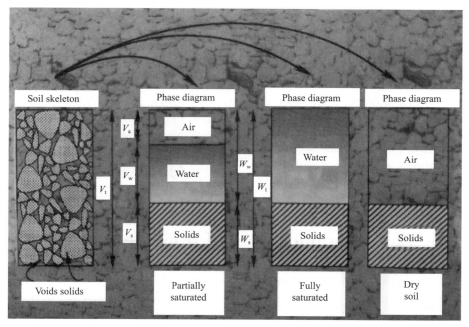

Fig. 12-1　Weight-Volume Relationships of Soil

12. 2　Soil Properties

In terms of engineering, soils differ in different locations because they are influenced by five soil formation factors. Those factors are the climate (kind of weather), vegetative cover (plants), parent material (bedrock), topography (sloping condition of the land) and time that the soil has had to form. The mineral portion of soil consists of sand, silt and clay particles. They differ in size, sand being the largest, silt being medium in size and clay being the smallest. Nearly all soils contain these three sizes, however the percentage of each varies tre-

mendously in different soils. The amounts of each of these particles determine the engineering properties and behavior of soil.

The first step of a large soil foundation-engineering project is a soil mechanics survey. The first visit on foot will show whether the site might be suitable. With respect to the behavior of soil, generally, the strength of soil increases with the depth in which it exists. But it can become weaker with depth. So, in choosing a foundation for this sort of soil, the knowledge of soil behavior and its engineering properties is essential, since this will give an idea of the likely settlements. There are, however, several other causes of settlement apart from consolidation due to load. These causes are incalculable and must be carefully guarded against. They include frost action, chemical change in the soil, underground erosion by flowing water, reduction of the ground water level and nearby construction of tunnels or vibrating machinery such as vehicles. Fig. 12-2 is the famous leaning tower of Pisa, which is caused by differential settlement. This figure shows that soil conditions need to be known in advance in order to properly support structure during its entire life time.

The engineering properties and behavior of soil are usually described by using soil profile. A soil profile generally consists of three main layers: the topsoil (100 ~ 200mm deep) or darker layer where air, water and humus allow plants to grow in; the sub-soil, a more clay-like layer which acts as a reservoir (water store) for the plants; the bedrock or parent material which is the underlying layer. Soil engineering property and behavior provide a relationship to structural properties such as strength, compressibility, permeability, swelling potential, etc.

Fig. 12-2   The Leaning Tower of Pisa, Italy

12.3   Applications of Soil Mechanics

Plasticity solutions are most commonly used based on the *Coulomb or Mohr-Coulomb failure*[4] conditions. Such solutions are used to examine the stability of a soil mass, that is, to determine the condition which will result in plastic collapse and to predict the mechanism of collapse. There are two ways in which the Coulomb failure criterion can be applied in stability, namely slip line and limit equilibrium methods respectively. Such solutions are appropriate in the following cases: (1) assuming that the failure criterion is satisfied at every point throughout a collapse zone; (2) assuming that collapse occurs as a result of the relative movement of rigid bodies and that plastic flow occurs only along the boundaries of such bodies.

In addition, based on *Darcy's Law*[5] and relative hydraulic theory, the permeability and seepage of soil can be studied, including the determination of coefficient of permeability, equation of continuity, flow nets, numerical analysis of seepage, safety of hydraulic structures against piping, etc. Another important part is consolidation settlement. By using consolidation theory, soil settlement can be calculated or evaluated.

Fig. 12-3 (a) and (b) are the famous buildings: Jinmao Tower and Oriental Pearl Tower. Both of them rest on soft soils with piling foundations in Shanghai. Foundations on Shanghai soils especially pile foundations, such as the foundations of Jinmao Tower and Oriental Pearl Tower, have won worldwide interest since the 1930s. During the process of design and construction, if wrong rules or measures are adopted, some damage may happen and sometimes it is fatal. Fig. 12-4 shows a landslide due to natural disasters, involving movement of large quantities of soils. We can clearly see the cracks caused by differential settlement in Fig. 12-5.

Fig. 12-3　Famous Buildings in Shanghai
(a) Jinmao Tower;(b) Oriental Pearl Tower

Fig. 12-4　Landslide

Fig. 12-5　Settlement Damage

12. 4　Foundations

To perform satisfactorily, every structure must have a proper foundation, namely a shallow foundation or deep foundation. When firm soil is not near the ground surface, a common shallow foundation for transferring the weight of a structure to the ground is through vertical members such as rafts, piles, piers or caissons. These terms do not have sharp definitions that distinguish one from another. Deep foundations are those where the soil support is applied at some depth below the usable portion of the structure.

Soil can vary considerably in its properties from one site or the other. In the same building site it can vary both in horizontal and vertical directions. For important structures, deep trial pits are dug so that the soil can be examined for some distance below the surface and sometimes loading tests on small areas at the bottom of the pits are made to estimate the safe load-bearing capacity of the soil. The purpose of a foundation is to convey the weight of a building to the soil in such a manner: (1) excessive settlement will not occur; (2) differential settlement of various parts of the building, which causes cracks in the structure will not occur; (3) the soil will not fail under its load thus causing no collapse of the building.

12. 4. 1   Categories of Foundations

Following is a list of some of the conditions that require different kinds of foundations: (1) Foundations of buildings and structures (Fig. 12-6); (2) Mat foundations (Fig. 12-7); (3) Retaining walls (Fig. 12-8); (4) Anchored sheet pile walls (Fig. 12-9); (5) Pile foundations (Fig. 12-10); (6) Pile groups (Fig. 12-11); (7) Raft and footings (Fig. 12-12); (8) Drilled pier (Fig. 12-13); (9) Caissons (Fig. 12-14).

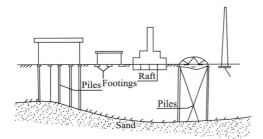

Fig. 12-6   Foundations of Buildings and Structures

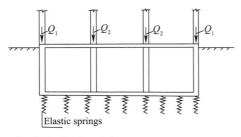

Fig. 12-7   Mat Foundations

12. 4. 2   Deep Foundations

With the development of economy in China, especially in big cities, such as Shanghai, Beijing and Guangzhou, deep foundations have become very popular to deal with soft soil. Generally, piles, caissons and piers are the most

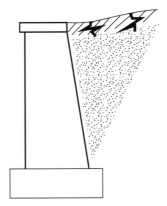

Fig. 12-8   Retaining Walls

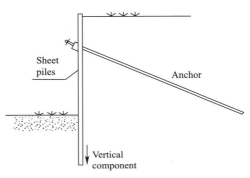

Fig. 12-9   Anchored Sheet Pile Walls

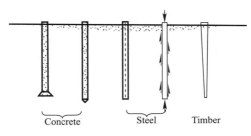

Fig. 12-10   Pile Foundations

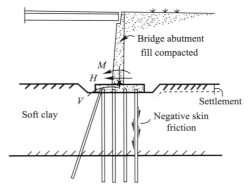

Fig. 12-11   Pile Groups

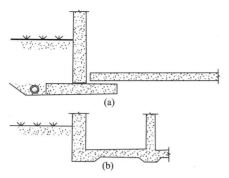

Fig. 12-12 Raft and Footings
(a) Footings;(b) Raft

important deep foundation types. Caissons and piers are larger in diameter than piles and are usually installed by driving.

（1）Pile foundations: Piles are structural members that are made of steel, concrete and/or timber. They are used to build pile foundations, which are deep and which cost more than shallow foundations. Despite the cost, the use of piles often becomes necessary to ensure that the structure under consideration is safe. Different types of pile are used in construction work depending on the type of load to be carried, the subsoil conditions and the ground water table. Piles can be divided into the following categories: （a）steel piles; （b）concrete piles; （c）wooden （timber）piles; （d）composite piles. Piles can be also divided into two categories: point bearing piles and friction piles.

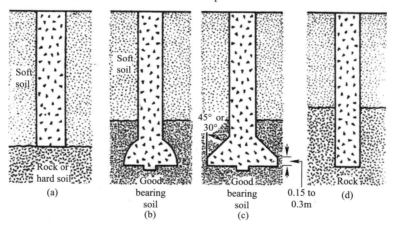

Fig. 12-13 Drilled Pier
(a) Straight-Shafted Pier; (b)、(c) Belled Pier; (d) Straight-Shafted Pier Rocketed into Rock

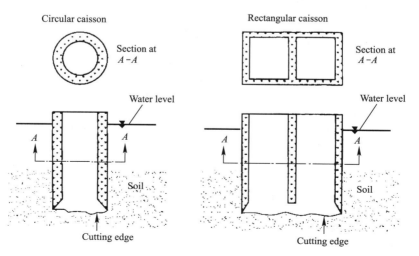

Fig. 12-14 Open Caissons

（2）Drilled pier：The terms caisson, pier, drilled shaft and drilled pier are often used interchangeably in foundation engineering; they all refer to a cast-in-place pile generally having a diameter of about 75mm or more, with or without steel reinforcements and with or without an enlarged bottom. Sometimes the diameter can be as small as 305mm. In order to avoid confusion, this text will use the following definitions: The term, drilled pier, is used when a hole is drilled or excavated to the bottom of a structure's foundation and then filled with concrete. *Depending on the soil conditions, casings or laggings (boards or sheet piles) may be used to prevent the soil around the hole from caving in during construction.* [6] The diameter of the pier shaft is usually large enough for a person to enter for inspection. According to the manner in which they are designed to transfer the structural load to the substratum, drilled piers can be classified into straight-shafted pier, belled pier and straight-shafted pier rocketed into rock (Fig. 12-13).

（3）Caissons：The term caisson refers to a substructure element used at wet construction sites, such as rivers, lakes and docks. For the construction of caissons, a hollow shaft or a box is sunk into position to rest on firm ground. The lower part of the shaft or the box is provided with a cutting edge to help it penetrate the soft soil layers below the water level and come to rest on a load-bearing stratum. The material inside the shaft or box is dredged through the openings at the top, and then concrete is poured in. Bridge abutments, quay walls and structures for shore protection can be built over caissons. Caissons can be divided into three major types: open caissons, box caissons and pneumatic caissons.

Vocabulary and Expressions：
pedologist [pi(:)'dɔlədʒist] n. 土壤学家
vicinity [və'sinəti] n. 邻近, 附近, 附近地区
in the vicinity of 在…附近; 在…上下
flocculate ['flɔkjuleit] v. 絮凝; (使)絮凝
mat foundation 筏式基础, 筏基, 整体基础

retaining wall 挡土墙
anchored sheet pile wall 锚定板桩墙
pile foundation 桩基
pile groups 群桩
drilled pier 钻孔桩
caisson 沉箱
forage ['fɔridʒ] n. 饲料, 草料, 搜索; vi. 觅食, 搜寻
humus ['hju:məs] n. 腐殖土, 腐殖质
feat [fi:t] n. 壮举, 伟业, 功绩, 技艺
auger boring 螺旋钻探
block resonant test 块体共振试验
caisson ['keisən] n. 沉箱
collapsible [kə'læpsəbəl] adj. 活动的, 可分解的, 可折叠的
dynamic penetration test 动力触探试验
earth-retainingstructures 挡土结构
field permeability test 现场渗透试验
friction pile 摩擦桩
group pile efficiency 群桩效率
group pile effect 群桩效应
in-situ 现场的
load-bearing capacity 承载力
peat [p:t] n. 泥炭, 泥煤, 泥炭土
pedestal ['pedistəl] n. 基座, 柱脚
mandrel ['mændrəl] n. 心轴
pier foundation 墩式基础
plate loading test 静力载荷试验
point bearing pile 端承桩
precast pile 预制桩
pressuremeter test 旁压试验
SPT(standard penetration test) 标准贯入度试验
static load test of pile 单桩静荷载试验
timber pile 木桩
vane shear test 十字板剪切试验
shale [ʃeil] n. [岩] 页岩; 泥板岩
misnomer [ˌmis'nəumə(r)] n. 用词不当, 误称
belled pier 扩底墩
box caisson 箱形沉箱
pneumatic caisson 气压沉箱

Notes：
① 太沙基(1948)对土力学的定义是："土力学是运用力学和水力学原理处理沉积物及其

他由岩石机械和化学分解作用而产生的松散固体颗粒沉积物的工程问题，不论沉积物是否包含有机混合物。"

② 从最大尺寸开始，大于 10cm 为大卵石，鹅卵石从 5~10cm，砾石或碎石大约从 5~50mm，砂砾大约从 2~5mm，砂从 0.06~2mm。由于所有这些土都很粗糙且没有黏性，所以颗粒之间都存在摩擦，其稳定性取决于颗粒之间的摩擦力。对于黏性土（或非摩擦土），两个主要的国际公认的尺寸范围是：粉砂（淤泥）从 0.002~0.02mm，而所有小于 0.002mm 的均属于黏土。

③ 表土结构的形成需要很长的时间，所以我们需要保护和保全土壤。一条经验法则是，一英寸（1 英寸=25.4 毫米）表土结构需要一百年的时间。一百年！请记住，正如我们今天需要良好的土壤一样，未来的人们也需要良好的土壤。最后一句 "A hundred years! Remember, just as we need good soil today, so will people in the years to come." 等同于 "A hundred years! Remember, just as we need good soil today, people in the years to come will need good soil too."

④ 莫尔失效或莫尔-库伦失效。

⑤ 达西定律。

⑥ 根据土壤条件，可以使用套管或垫板（板桩）来防止在施工过程中桩周土的塌陷。

Exercises：

1. Answer the Following Questions in English.

（1）In your own words, explain the reasons underlying the definition of "soil mechanics and foundations" given in this chapter.

（2）Using examples, explain the concept of different kinds of foundations.

（3）Comment on "soil types".

（4）How does soil mechanics and foundation differ?

（5）Assuming that you are a civil engineer, how could society's demands for improvement in the quality of life affect your attitudes and practices?

（6）Mention some recent governmental laws or regulations that can affect civil engineering in the future.

2. Translate the Following Sentences into Chinese.

（1）The larger the all-around confining pressure, the greater the shear strength, and also the greater the shear modulus G.

（2）On the other hand, if the stresses are increased very slowly, the water is forced out of the voids and this is the drained condition.

（3）The pore pressure u_b for an unsaturated soil is a combination of the pore air and porewater pressures.

（4）It is thus seen that a change in porewater pressure always accompanies a stress change and that its magnitude is governed by the compressibility and expansibility of the soil.

3. Translate the Following Sentences into English.

（1）深基础是连接建筑物上部结构并将其荷载直接传至地表以下较深土层或岩层的基础，通常基础埋深超过 5m 或不小于基础宽度，一般用于荷载较大、对地基承载力和变形要求较高的工程。

（2）原位测试是在现场原位基本保持土的天然结构、天然含水量及天然应力状态下测定土的力学性能的试验。

（3）地下水水位及其变化情况的调查以及渗透系数的确定是场地勘察的必要组成部分。

（4）静载荷试验可以用来评价地基的承载力与沉降特征。

（5）可以看出随着土中应力的增加，土会变得更硬。

（6）如果应力施加得很快，则加荷后水不会立即从土中排出。

（7）基础的埋深要有足够的深度，以避免由于一些影响因素而引起基础的过大移动。

（8）沉降量的确定方法和容许值的大小将在另一节中讨论。

13 Masonry Structures
（砌体结构）

教学目标：本单元的主要内容有砌体、砌块、砌体结构分类、砌体墙类型以及建筑规范限值。教学重点是砌体结构分类，难点是建筑规范限值。通过本单元的学习，要求熟悉砌体、砌块和砌体墙类型，掌握砌体结构和砌体墙的分类。

13.1　Introduction

Masonry is one of man's oldest building materials and probably one of the most maligned and most certainly the least understood. [①] Perhaps because of the considerable amount of information and data available today, both as to its properties and structural performance, sound design techniques and vastly improved construction practices have evolved within recent years, all of which make for optimum use of the material's capabilities. This is in no small way due to the effort continually being exerted toward this evolution by such diverse agencies as the International Conference of Building Officials (ICBO) and the Masonry Institute of America (MIA). Furthermore, the wind, seismic, and structural performance research carried on during the recent past has resulted in building codes of increasing complexity. This, in turn, has led to more sophisticated and comprehensive methods of design.

Masonry is primarily a hand-placed material whose performance is highly influenced by factors of placement. Hence, knowledge of the basic ingredients (i.e., mortar, grout, masonry unit and reinforcement) is essential if a practical and efficient design conception is to be achieved. In addition, if the design is to be brought to a successful fruition, as its designer conceived it, proper inspection procedures must be followed to ensure that its delivery will be more certain. Furthermore, before anyone can hope to turn out an adequate design of any sort, he or she must possess a rudimentary knowledge of the properties and performance of the materials being employed. The total design of a masonry building begins with a consideration of the preliminary and nonstructural aspects of masonry bearing on the case study, such as its fire-resistive or environmental features. Following this examination come the determination of the live, dead, seismic and wind loads—their magnitudes and stress paths from point of application to the ground. Finally, the member sizes and reinforcing requirements are selected, adequate connections are devised, and the system is detailed such that it can be readily constructed.

Brick is actually the oldest manufactured building material remaining in use today. In the premodern era, the development of brick masonry reached its fruition in the United States and Europe. The successful use of this ancient material is certainly demonstrated in many early American brick structures, such as the Monadnock Building in Chicago. But its very massiveness discouraged further use of unreinforced masonry bearing walls for high-rise buildings.

13.2　Classifications of Masonry Construction

The classifications of masonry construction and the types of masonry walls appear in the *UBC*. [②] The distinction between these various categories must be thoroughly understood by anyone who intends to design masonry under UBC jurisdiction. For this reason, they are thoroughly delineated in the following sections.

Masonry construction is classified as follows: (1) "Reinforced masonry", which must be engineered on the basis of sound theoretical principles combined with a set of empirical rules and limitations set forth by the Building Code, plus sound engineering judgment stemming from long experience; (2) "Partially reinforced masonry", which is introduced into the Uniform Building Code primarily for those areas in which all the requirements of reinforced masonry are not needed, since the seismicity of the locale does not so dictate; (3) "Unreinforced engi-neered masonry", which is developed in the East as an attempt to improve on past practices, many of which are unsound; (4) "Traditional masonry", which encompasses the use of masonry as it evolved over the years from certain arbitrary limitations and past practices without any real consideration for theoretical design characteristics, although it does provide for a generally conservative and safe type of construction for the majority of conditions.

13.3　Types of Masonry Walls

(1) Unburned clay masonry: Unburned clay masonry consists of unburned clay units, commonly refers to as "adobe" in the south-western part of the United States. In earlier and less sophisticated days, it did perform quite sat-isfactorily where no seismic activity of any mag-nitude occurred. The early adobe is actually re-inforced with straw and often also contains an emulsion that provides for greater compressive strength and durability. The church located in the Los Angeles Plaza, is built of sun-dried clay and protected by plaster.

(2) Gypsum masonry: Gypsum masonry consists of gypsum block or gypsum tile units laid up with gypsum mortar. It has been used in the past, with considerable success, for interior partitions, primarily because of the ease with which it can be formed around ducts, window opening and other discontinuities. It is also per-mitted in some "partially reinforced" walls. Gypsum tile is laid up in gypsum mortar, simi-lar to that used in plaster. The proportions con-sist of approximately one part gypsum to three parts sand, mixed with a sufficient amount of water to provide a good workable mortar. Since gypsum is fast-setting, the mortar sets up so rapidly that it has a limited "board" life. Thus, it is generally necessary to add a retarder of some sort, composed usually of certain organic materials. Gypsum tile, like unreinforced brick masonry, *received a bad press*[3] after the 1933 Long Beach Earthquake, again because of mate-rial misuse, not because of any property defi-ciencies.

(3) Glass masonry: Glass masonry units are used in the openings of non-load-bearing ex-terior or interior walls. These filler panels must be at least 76.2mm thick and the mortared sur-faces of the blocks have to be treated to provide an adequate mortar-bonding effect. This is usu-ally achieved by applying a roughened surface abhesive to the glass edges. The panels them-selves must be restrained laterally to resist the lateral-force effects of winds or earthquakes. Reinforcement, as required by calculations, is provided. Exterior glass block panels have to be provided with 12.7mm expansion joints at the sides and at the top, and they must be entirely free of mortar so that the space can be filled with a resilient material to provide for needed movement. The expansion joint, of course, must also provide for lateral support while per-mitting expansion and contraction of the glass panel.

(4) Stone masonry: Stone masonry is that form of construction made with natural or cast stone as the basic masonry unit, set in mortar with the joints thoroughly filled. In ashlar ma-sonry, the bond stones are uniformly distributed and have to cover at least 10% of the area of the

exposed facets. Rubble stone masonry, 609.6mm or less in thickness, will have bond stones spaced a maximum of 914.4mm both vertically and horizontally. Should the thickness exceed 609.6mm, the bond stone spacing is increased to 1.8288m on both sides. There are other limits, arbitrarily established. The maximum height/thickness ratio is 14, and the minimum wall thickness is 406.4mm. If regularly cut or shaped stones are used, they may be laid as solid or grouted brick masonry.

(5) Cavity wall masonry: Cavity wall masonry is construction using brick, structural clay tile, concrete masonry or any combination thereof, in which the facing and the backing wythes are completely separated except for metal ties that serve as cross ties or bonding elements. The maximum height/thickness ratio is limited to 18, with the minimum thickness being 203.2mm. The cavity wall facing and backing wythes cannot be less than 101.6mm in thickness, except that when both are constructed with clay or shale brick the limit decreases to 76.2mm nominal thickness. The two wythes have to be bonded together with 4.7625mm metal ties embedded in the horizontal mortar joints. Tie spacing is limited such that they support no more than $0.418m^2$ of wall area for cavity widths up to 88.9mm. Additional bonding ties must be placed at all openings, spaced at 914.4mm maximum around the perimeter of the openings, within 304.8mm of the openings.

(6) Hollow unit masonry: Hollow unit masonry describes a type of wall construction that consists of hollow masonry units set in mortar as they are laid in the wall. All units have to be laid with full-face shell mortar beds, with the head or end joints filled solidly with mortar for a distance in from the face of the unit not less than the thickness of the longitudinal face shells. This type of construction usually refers to an unreinforced state, although it actually can be reinforced. Where the wall thickness consists of two or more hollow units placed side by side, the stretcher unit must be bonded at vertical intervals not to exceed 863.6mm. This bonding is accomplished by lapping a block at least

101.6mm over the unit below, or by lapping them at vertical intervals not to exceed 431.8mm with units that are at least 50% greater in thickness than the units below. They can also be bonded together with corrosion-resistant metal ties which conform to those requirements for cavity walls, as previously noted. Since this material is not reinforced, the maximum height/thickness ratio is 18, with a minimum thickness of 203.2mm.

(7) Solid masonry: Solid masonry consists of brick, concrete brick or solid load-bearing concrete masonry units laid up contiguously in mortar. All units are laid with full shoved mortar joints, and the head, bed and wall joints have to be solidly filled with mortar. In each wythe, at least 75% of the units in any vertical transverse plane must lap the ends of the unit above and below a distance not less than 38.1mm, nor less than one-half the height of the units, whichever is greater. Otherwise, the masonry is to be reinforced longitudinally to provide for a loss of bond, as in the case of masonry laid in stack bond. *The longitudinal reinforcement amounts to a minimum of two continuous wires in each wythe, with a minimum total cross-sectional area of $10.968mm^2$, being provided in the horizontal bed joints, with the spacing not to exceed 406.4mm center to center vertically.* ④

(8) Grouted masonry and reinforced grouted masonry: Grouted masonry is made with two wythes of clay brick or solid concrete brick or stone units in which the interior joint, sometimes called the "collar joint" or interior wythe is filled with grout. This bonds the two wythes together as well as providing a space wherein the reinforcement can be located and bonded to the surrounding masonry. The thickness of grout or mortar between masonry units and reinforcement is not to be less than 25.4mm, except that steel wire reinforcement may be laid in horizontal mortar joints which are at least twice the thickness of the wire diameter.

Because of the poor performance of unreinforced brickwork, kilns are closed and the brick industry languished. *However, the structural en-*

gineering profession, *through its professional society (Structural Engineers Association of California), has awakened to its responsibility to the public, and in conjunction with the then California State Division of Architecture has developed reinforced grouted bonded walls, initially for school masonry construction.* [5] As this school building type of construction improves and evolves, grouting becomes recognized as a means of overcoming the inherent tensile weakness in the material by providing a superior bond between reinforcing and brickwork, making it a more homogeneous material. By so doing, it permits the wall or beam element to readily resist flexural tensile stresses of considerable magnitude.

(9) Reinforced hollow unit masonry: Reinforced hollow unit masonry is laid up as described in the UBC, but the reinforcing is grouted within the vertical cells, and horizontal reinforcing consists of joint reinforcing in the bed joints or of horizontal steel placed in the bond beam units. This method is originally developed for reinforced concrete block walls, but it is now also used in reinforced hollow clay brick construction.

(10) Masonry veneer: Masonry veneer is a nonstructural installation of facing material or decorative surfacing attached to a previously constructed structural element. Two basic types exist presently, adhered and anchored veneers.

13.4　Empirical Code Limits

Building codes of one sort or another have been developing worldwide, probably even before Hammurabi's time, perhaps to protect the designer's health. For in those days, designers lost their lives if one of their structures collapsed and caused someone's death. To a lesser degree, that practice has continued to the present day. In some countries, Greece, for example, should a serious structural collapse occur, the designer or project engineer is jailed until liability has been established. Most of these empirical relations may not have theoretical basis in fact, but they produce results that work. *They have evolved from a "cut-and-try" process over the years. Now we call the process "successive approximation" or "iteration". These terms sound more sophisticated, and besides they are less informative—a characteristic of much of our modern technological language. Call it what you will, a successive approximation is no more scientific than a cut-and-try approach.* [6] As an example of such limits, note that the height/thickness ratio (h/t) of masonry walls is limited to an arbitrarily selected ratio. Consider, for example, the h/t limit of 14 for unreinforced stonework. When building the old city walls of Rothenburg, the primitive masons on the project somehow simply sensed that this was a safe limit to be imposed. *They knew virtually nothing about material strength or any conditions of wall restraint. But, it turns out, this is a very reasonable limit, albeit a conservative one.* [7]

Vocabulary and Expressions：

abhesive [æb'hi：siv] n. 阻黏剂
accentuate [æk'sentjueit] v. 重读，强调
adobe [ə'dəubi，ə'dəub] n. 砖坯，土砖
ashlar ['æʃlə] n. 方石堆，装饰屋内墙面的石板
bed joint 底层接缝，平缝，平层节理
bolster ['bəulst] vt. 支持，支援

bond beam 结合梁，圈梁
bond stone 砌合石
brickwork ['brikwə：k] n. 砌砖
cavity wall 空心墙
chagrin ['ʃægrin] n. 懊恼，气愤，委屈
chicken-wire 铁丝织网，细号钢丝网
cut-and-try 试验性的

dead load 恒荷载

diaphragm ['daiəfræm] n. 横隔板，横隔墙

disrepute [disri'pjuːt] n. 坏名声，无信誉

emulsion [i'mʌlʃən] n. 乳状液，乳胶

engineer [endʒi'niə] vt. 设计，策划

face shell（空心块材的）外壁

fruition [fruː'iʃən] n. 成就，实现

gypsum ['dʒipsəm] n. 石膏

head（end）joint 端灰缝

header ['hedə] n. 顶砖

height/thickness ratio 高厚比

high-lift 高压的，高扬程的

hollow unit masonry 空心块材砌体

languish ['læŋgwiʃ] vi. 衰退

liability [laiə'biliti] n. 责任，义务，债务，负债

live load 活荷载

locale [英 ləu'kaːl，美 ləu'kæl] n. 现场，场所

low-lift 低压的，低扬程的

malign [mə'lain] adj. 有害（的），不良（的）

masonry unit（砌体的）块材

misconception [miskən'sepʃən] n. 误解

mortar bed 砂浆平缝

mortarjoint 灰缝

ostensibly [ɔs'tensibli] adv. 表面上，假装地

partition [paː'tiʃən] n. 分隔墙

precept ['priːsept] n. 规则

resilient [ri'ziliənt] adj. 弹性的，有回弹力的

retarder [ri'taːdə] n. 缓凝剂

rubble ['rʌbəl] n. 毛石，块石，碎石

set [set] vi. 凝固

shale [ʃeil] n. 页岩，泥板岩

shear wall 剪力墙

shoved mortarjoint 挤浆砌筑的灰缝

stack bond 竖向通缝砌筑

stretcher unit 顺砌块材

substantiate [səb'stænʃieit] vt. 证实，证明

veneer [və'niə] n. 饰面，镶板，饰面砖，墙面砖

watershed ['wɔːtəʃed] n. 分水岭，流域，汇水区

wythe [wiθ] n.（建筑中）砖的厚度

in lieu [ljuː/luː] of 代，代替

albeit [ˌɔːl'biːit] conj. 虽然，尽管，固

然，即便

Notes：

① 砌体是人类最古老的建筑材料之一，也可能是最受诋毁、最不为人所了解的材料之一。"the most maligned"和"the least understood"并列对应，least 是 little 的最高级；副词短语"most certainly"修饰"the least understood"。

② UBC：Uniform Building Code 统一建筑规范。

③ 收到各报刊的不好的评价。

④ 水平砖缝中至少要配有两根连续的纵向加固钢筋（钢筋最小总横截面面积为10.968mm²），相邻两条配有纵向加固钢筋的水平砖缝间的垂直间距（砖缝中心线到砖缝中心线）不超过406.4mm。

⑤ 然而，结构工程专业通过其专业协会（加州结构工程师协会）意识到其对公众的责任，并与当时的加州建筑部门合作开发了钢筋灌浆黏结墙，最初用于学校砌体建筑。句中"then"的意思是"当时的"。

⑥ 这些年来，它们是由"试一试"的过程演变而来的。现在我们称这个过程为"逐次逼近"或"迭代"。这些术语听起来更复杂，而且它们的信息量更少——这是我们现代技术语言的一个特征。不管你怎么称呼它，逐次逼近的方法并不比试一试的方法更科学。

⑦ 他们对材料强度和任何墙体约束条件几乎一无所知。但事实证明，虽然它是一个保守的限值，但它确实是一个非常合理的限值。

Exercises：

1. Distinguish betweenHollow Unit Masonry and Solid Masonry.

2. Describe Grouted Masonry and Reinforced Grouted masonry.

3. Translate the Subsection "Alernative Materials and Construction Method" into Chinese.

4. Translate the Following Paragraphs into Chinese.

（1）The Slenderness Ratio, *SR*

$$SR = \frac{\text{effective height(or length)}}{\text{effective thickness}}$$

　　Whichever is the lesser—the effective

height or effective length—is used to calculate the slenderness ratio. Obviously as the slenderness ratio increases the loadbearing capacity decreases.

(2) Wind load is imposed on the external faces of buildings as horizontal loads—pressures or suctions. The effect of wind on a building can also develop internal pressures and suctions, and these are additive to the external values. The wind pressure is affected by a number of factors, such as building size, shape and location, and it generally increases with height.

5. Translate the Following Sentences into English.

(1) 对于砖砌体，砖的含水率较大时易于保证砌筑质量，干砖砌筑和用含水饱和的砖砌筑都会降低砖与砂浆的黏结强度，从而降低砌体的抗压强度。

(2) 砌体的高厚比类似于钢筋混凝土结构构件中的长细比。

(3) 在截面尺寸和砌体材料强度等级一定的条件下，影响砌体受压构件承载力的主要因素是构件的高厚比和轴向力的偏心距。

14　Reinforced Concrete and Prestressed Concrete （钢筋混凝土与预应力混凝土）

教学目标：本单元的主要内容有钢筋、现浇混凝土、预制混凝土、素混凝土、钢筋混凝土性能及其优缺点、恒荷载、活荷载、钢筋混凝土与预应力混凝土的比较以及施加预应力的方法（先张法、后张法）。教学重点是钢筋混凝土性能及其优缺点，难点是施加预应力的方法。通过本单元的学习，要求熟悉现浇混凝土、预制混凝土、素混凝土、钢筋混凝土、恒荷载和活荷载的概念，掌握钢筋混凝土性能及其优缺点，了解施加预应力的方法（先张法、后张法）。

14.1   Reinforced Concrete

Concrete and reinforced concrete are used as building materials in every country. In many, including the United States and Canada, reinforced concrete is a dominant structural material in engineered construction. The universal nature of reinforced concrete construction stems from the wide availability of *reinforcing bars*[1] and the constituents of concrete, gravel, sand and cement, the relatively simple skills required in concrete construction, and the economy of reinforced concrete compared to other forms of construction. Concrete and reinforced concrete are used in bridges, buildings of all sorts, underground structures, water tanks, television towers, offshore oil exploration and production structures, dams and even in ships.

Reinforced concrete structures may be *cast-in-place concrete*[2], constructed in their final location, or they may be *precast concrete*[3] produced in a factory and erected at the construction site. Concrete structures may be severe and functional in design, or the shape and layout can be whimsical and artistic. Few other building materials offer the architect and engineer such versatility and scope. Concrete is strong in compression but weak in tension. As a result, cracks develop whenever loads, or restrained shrinkage or temperature changes, give rise to tensile stresses in excess of the tensile strength of the concrete. In *a plain concrete beam*[4], the moments about the neutral axis due to applied loads are resisted by an internal tension-compression couple involving tension in the concrete. Such a beam fails very suddenly and completely when the first crack forms. In a reinforced concrete beam, steel bars are embedded in the concrete in such a way that the tension forces needed for moment equilibrium after the concrete cracks can be developed in the bars.

The construction of *a reinforced concrete member*[5] involves building a form or mold in the shape of the member being built. The form must be strong enough to support both the weight and hydrostatic pressure of the wet concrete, and any forces applied to it by workers, concrete buggies, wind and so on. The reinforcement is placed in this form and held in place during the concreting operation. After the concrete has hardened, the forms are removed. As the forms are removed, props or shores are installed to support the weight of the concrete until it has reached sufficient strength to support the loads by itself. The choice of structural system is made by the architect or engineer early in the design, based on the following considerations:

（1）Economy: Frequently, the foremost consideration is the overall cost of the structure. This is, of course, a function of the costs of the materials and the labor necessary to erect them. Frequently, however, the overall cost is affected as much or more by the overall construction time since the contractor and owner must borrow or otherwise allocate money to carry out the construction and will not receive a return on this investment until the building is ready for occupancy. In a typical large apartment or commercial project, the cost of construction financing will be a significant fraction of the total cost. As a result, financial savings due to rapid construction may be more than offsetting increased material costs. For this reason, any measures which the designer can take to standardize the design and forming will generally pay off in reduced overall costs. In many cases the long-term economy of the structure may be more important than the first cost. As a result, maintenance and durability are important consideration.

(2) Suitability of material for architectural and structural function: A reinforced concrete system frequently allows the designer to combine the architectural and structural functions. Concrete has the advantage that it is placed in a plastic condition and is given the desired shape and texture by means of the forms and the finishing techniques. This allows such elements as flat plates or other types of slabs to serve as load-bearing elements while providing the finished floor and/or ceiling surfaces. Similarly, reinforced concrete walls can provide architecturally attractive surfaces in addition to having the ability to resist gravity, wind or seismic loads. Finally, the choice of size or shape is governed by the designer and not by the availability of standard manufactured members.

(3) Fire resistance: The structure in a building must withstand the effects of a fire and remain standing while the building is evacuated and the fire is extinguished. A concrete building inherently has a 1- to 3-hour fire rating without special fireproofing or other details. Structural steel or timber buildings must be fireproofed to attain similar fire ratings.

(4) Low maintenance: Concrete members inherently require less maintenance than do structural steel or timber members. This is particularly true if dense, air-entrained concrete has been used for surfaces exposed to the atmosphere, and if care has been taken in the design to provide adequate drainage off and away from the structure. Special precautions must be taken for concrete exposed to salts such as deicing chemicals.

(5) Availability of materials: Sand, gravel, cement and concrete mixing facilities are very widely available, and reinforcing steel can be transported to most job sites more easily than structural steel. As a result, reinforced concrete

is frequently used in remote areas. On the other hand, there are a number of factors that may cause one to select a material other than reinforced concrete. These include:

(a) Low tensile strength: The tensile strength of concrete is much lower than its compressive strength (about 1/10), and hence concrete is subject to cracking. In structural uses this is overcome by using reinforcement to carry tensile forces and limit crack widths to within acceptable values. Unless care is taken in design and construction, however, these cracks may be unsightly or may allow penetration of water. When this occurs, water or chemicals such as road deicing salts may cause deterioration or staining of the concrete. Special design details are required in such cases. In the case of water-retaining structures, special details and/or prestressing are required to prevent leakage.

(b) Forms and shoring: The construction of a cast-in-place structure involves three steps not encountered in the construction of steel or timber structures. These are the construction of the forms, the removal of these forms and propping or shoring the new concrete to support its weight until its strength is adequate. Each of these steps involves labor and/or materials, which are not necessary with other forms of construction.

(c) Relatively low strength per unit of weight or volume: The compressive strength of concrete is roughly 5% to 10% that of steel, while its unit density is roughly 30% that of steel. As a result, a concrete structure requires a larger volume and a greater weight of material than does a comparable steel structure. As a result, long-span structures are often built from steel.

(d) Time-dependent volume changes: Both concrete and steel undergo approximately the same amount of thermal expansion and contraction. Because there is less mass of steel to be heated or cooled, and because steel is a better conductor than concrete, a steel structure is generally affected by temperature changes to a greater extent than is a concrete structure. On the other hand, concrete undergoes drying shrinkage, which, if restrained, may cause deflections or cracking. Furthermore, deflections will tend to increase with time, possibly doubling, due to creep of the concrete under sustained loads.

Since reinforced concrete isa nonhomogeneous material that creeps, shrinks and cracks, its stresses cannot be accurately predicted by the traditional equations derived in a course in strength of materials for homogeneous elastic materials. Much of reinforced concrete design is therefore empirical; i.e., design equations and design methods are based on experimental and time-proved results instead of being derived exclusively from theoretical formulations. A thorough understanding of the behavior of reinforced concrete will allow the designer to convert an otherwise brittle material into tough ductile structural elements and thereby take advantage of concrete's desirable characteristics, its high compressive strength, its fire resistance and its durability.

Concrete, a stonelike material, is made by mixing cement, water, fine aggregate (often sand), coarse aggregate and frequently other additives (that modify properties) into a workable mixture. In its unhardened or plastic state, concrete can be placed in forms to produce a large variety of structural elements. Although the hardened concrete by itself, i.e., without any reinforcement, is strong in compression, it lacks tensile strength and therefore cracks easily. Because unreinforced concrete is brittle, it cannot undergo large deformations under load and fails suddenly without warning. The addition of steel reinforcement to the concrete reduces the negative effects of its two principal inherent weaknesses, its susceptibility to cracking and its brittleness. When the reinforcement is strongly bonded to the concrete, a strong, stiff and ductile construction material is produced. This material, called reinforced concrete, is used extensively to construct foundations, structural frames, storage tanks, shell roofs, highways, walls, dams, canals and innumerable other structures and building products. Two other characteristics of concrete that are present even when concrete is reinforced are shrinkage and creep, but the negative effects of these properties can be mitigated by careful design.

The load associated with the weight of the structure itself and its permanent components is called the dead load. The dead load of concrete members, which is substantial, should never be neglected in design computations. The exact magnitude of the dead load is not known accurately until members have been sized. Since some figure for the dead load must be used in computations to size the members, its magnitude must be estimated at first. After a structure has been analyzed, the members have been sized and architectural details have been completed, the dead load can be computed more accurately. If the computed dead load is approximately equal to the initial estimate of its value (or slightly less), the design is complete, but if a significant difference exists between the computed and estimated values of dead weight, the computations should be revised using an improved value of dead load. An accurate estimate of dead load is particularly important when spans are long, say over 75ft (22.9m), because dead load constitutes a major portion of the design load.

Live loads associated with building use are specified by city or state building codes. Instead of attempting to evaluate the weight of specific items of equipment and occupants in a certain area of a building, building codes specify values of uniform live load for which members are to be designed. After the structure has been sized for vertical load, it is checked for wind in combination with dead and live load as specified in the code. Wind loads do not usually control the size of members in buildings less than 16 to 18 storeys, but for tall buildings wind loads become significant and cause large forces to develop in the structures. Under these conditions economy can be achieved only by selecting a structural system that is able to transfer horizontal loads into the ground efficiently.

14.2   Prestressed Concrete

Concrete is strong in compression, but weak in tension: its tensile strength varies from 8% to 14% of its compressive strength. Due to such a low tensile capacity, flexural cracks develop at early stages of loading. In order to reduce or prevent such cracks from developing, a concentric or eccentric force is imposed in the longitudinal direction of the structural element. This force prevents the cracks from developing by eliminating or considerably reducing the tensile stresses at the critical midspan and support sections at service load, thereby raising the bending, shear and torsional capacities of the sections. The sections are then able to behave elastically, and almost the full capacity of the concrete in compression can be efficiently utilized across the entire depth of the concrete sections when all loads act on the structure.

Such an imposed longitudinal force is calleda prestressing force, i.e., a compressive force that prestresses the sections along the span of the structural element prior to the application of the transverse gravity, dead and live loads or transient horizontal live loads. The type of prestressing force involved, together with its magnitude, are determined mainly on the basis of the type of system to be constructed and the span length and *slenderness*[⑥] desired. Since the prestressing force is applied longitudinally along or parallel to the axis of the member, the prestressing principle involved is commonly known as linear prestressing. Circular prestressing, used in liquid containment tanks, pipes and pressure reactor vessels, essentially follows the same basic principles, as does linear prestressing. The circumferential hoop, or "hugging" stress on the cylindrical or spherical structure, neutralizes the tensile stresses at the outer fibers of the curvilinear surface caused by the internal contained pressure.

Fig. 14-1 illustrates, in a basic fashion, the prestressing action in both types of structural systems and the resulting stress response. In Fig. 14-1(a), the individual concrete blocks act together as a beam due to the large compressive prestressing force P. Although it might appear that the blocks would slip and vertically simulate shear slip failure, in fact they will not because of the longitudinal force P. Similarly, the wooden staves in Fig. 14-1(c) might appear to be capable of the separating as a result of the high internal radial pressure exerted on them. But again, because of the compressive prestress imposed by the metal bands as a form of circular prestressing, they will remain in place.

14.2.1   Comparison with Reinforced Concrete

From the preceding discussion, it is plain that permanent stresses in the prestressed structural member are created before the full dead and live loads are applied in order to eliminate or considerably reduce the net tensile stresses caused by these loads. With reinforced concrete, it is assumed that the tensile strength of the concrete is negligible and disregarded. This is because the tensile forces resulting from the bending moments are resisted by the bond created in the reinforcement process. Cracking and deflection are therefore essentially irrecoverable in reinforced concrete once the member has reached its limit state at service load.

The reinforcement in the reinforced concrete member does not exert any force of its own on the member, contrary to the action of prestressing steel. The steel required to produce the prestressing force in the prestressed member actively preloads the member, permitting a relatively high controlled recovery of cracking and deflection. Once the flexural tensile strength of the concrete is exceeded, the prestressed member

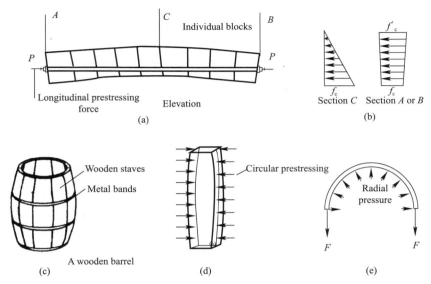

Fig. 14-1   Prestreesing Principle in Linear and Circular Prestressing

(a) Linear Prestressing of a Series of Blocks to Form a Beam; (b) Compressive tress on Midspan Section C and End Section A or B;

(c) Circular Prestressing of a Wooden Barrel by Tensioning the Metal Bands; (d) Circular Hoop Prestress on One Wooden Stave;

(e) Tension Force F on Half of Metal Band due to Internal Pressure to Be Balanced by Circular Hoop Prestress

starts to act like a reinforced concrete element. By controlling the amount of prestress, a structural system can be made either flexible or without influencing its strength. In reinforced concrete, such a flexibility in behavior is considerably more difficult to achieve if considerations of economy are to be observed in the design. Flexible structures such as fender piles in wharves have to be highly energy absorbent, and prestressed concrete can provide the required resiliency. Structures designed to withstand heavy vibrations, such as machine foundations, can easily be made rigid through the contribution of the prestressing force to the reduction of their otherwise flexible deformation behavior.

14. 2. 2   Economics

Prestressed members are shallower in depth than their reinforced concrete counterparts for the same span and loading conditions. In general, the depth of a prestressed concrete member is usually about 65% to 80% of the depth of the equivalent reinforced concrete member. Hence, the prestressed member requires less concrete and about 20% to 35% of the amount of reinforcement. Unfortunately, this saving in material weight is balanced by the higher cost of the higher quality materials needed in prestressing. Also, regardless of the system used, prestressing operations themselves result in an added cost: formwork is more complex, since the geometry of prestressed sections is usually composed of flanged sections with thin webs.

In spite of these additional costs, if a large enough number of precast units are manufactured, the difference between at least the initial costs of prestressed and reinforced concrete systems is usually not very large. And the indirect long-term savings are quite substantial, because less maintenance is needed, a longer working life is possible due to better quality control of the concrete, and lighter foundations are achieved due to the smaller cumulative weight of the superstructure. Once the beam span of reinforced concrete exceeds 21. 336 to 27. 432m, the dead weight of the beam becomes excessive, resulting in heavier members and, consequently, greater long-term deflection and cracking. Thus, for larger spans, prestressed concrete becomes mandatory since arches are expensive to construct and do not perform as well due to the severe long-term shrinkage and creep they undergo. Very large spans such as segmental bridges or cable-stayed bridges can only be constructed through the use of prestressing.

14. 2. 3　Prestressing Methods

There are three principal ways in which concrete is prestressed, namely, by tensioned wires fixed at their ends, by tensioned wires gripped by the concrete and by applying an external load; experiments are also being made with expanding-cement concrete.

End-anchored wires are used in long-span concrete beams. The reinforcement is separated from the surrounding concrete by a sheath or other wrapping, which allows the wires to move freely during stressing. The wires are anchored at the ends of the beam either by fixing to an anchor plate, or by other devices such as wedging with concrete cones. Since the steel is stressed after the concrete is cast and hardened, this method of prestressing is known as *post-tensioning*[7] (Fig. 14-2), A disadvantage of this method is that the wires must be protected from corrosion by forcing cement grout into the sheath. The grout provides protection against corrosion and also provides a bond between the wires and the sheath and thus with the concrete; it also supplements the resistance of the wires to slip, without which the security of the wires would depend entirely on the permanence of the end anchorages. With proper care these disadvantages do not involve serious risk. The object of allowing the wires to move while the prestress is being established is to make it possible to prestress the beam after most of the shrinkage of the concrete has taken place, and so reduce the loss of prestress due to shrinkage and to eliminate loss of prestress due to elastic contraction of the concrete. In

some cases, it is possible to increase the prestress when the beam begins to carry its own weight, the prestress thus relieving the member of a considerable proportion of compressive stress due to its weight.

In the concrete-gripped type of prestressed concrete, the wires are stretched before concreting; the method is consequently referred to as *pre-tensioning*[8] (Fig. 14-3). When it has hardened the concrete grips the wires as in reinforced concrete, except that the grip may be increased slightly when the wires are released from the stretching device on account of a slight shortening and swelling of the wires that occur as the concrete member shortens under compression. The shortening of the concrete member is also a slight disadvantage of this method of prestressing, since the consequent shortening of the tensioned wires is accompanied by a reduction in the prestress, in addition to that due to shrinkage of the concrete. On the other hand it is not necessary to provide anti-corrosive treatment, the wires are gripped throughout their length, and the security of the beam is not dependent upon anchorage of the wires at the ends. Prestressing may increase the ultimate moment of resistance of the concrete in a beam; failure generally occurs by the yielding of the wires. If, therefore, sufficient prestress can be induced by this method to eliminate cracking under a small overload, it is probably better than the post-tensioning method, since there is no anchorage to fail. The application of the post-tensioning method is often simpler to carry out on the site than pre-tensioning.

The use of expanding-cement concrete for inducing a prestress is introduced by M. Lossier and is still in the experimental stage. The prestress is caused by the concrete swelling instead of shrinking during setting and hardening. If this process can be developed so that the initial prestresses are sufficiently high to prevent cracking occurring under load, it may become a useful method of construction.

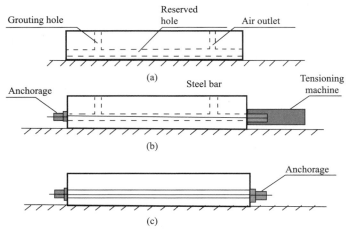

Fig. 14-2   Post-Tensioning Method
(a) Concrete Member; (b) Post-Tensioning; (c) Anchored by Anchorage and Grouting the Hole

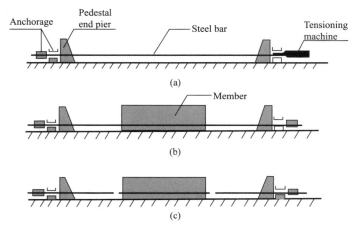

Fig. 14-3   Pre-Tensioning Method
(a) Pre-Tensioning; (b) Forming and Placement; (c) Cutting Steel Bars

Vocabulary and Expressions：

form［fɔːm］n. 模板

buggy［'bʌgi］n. 小推车

shore［ʃɔː(r)］n. 斜撑，斜撑柱

deice［diː'ais］v. 除冰，防冻

props or shores 支撑

creep［kriːp］vi. n. 徐变

eccentric［ik'sentrik］adj. 偏心的

resiliency［ri'ziliənsi］n. 弹性

transient［'trænziənt］adj. 短暂的，瞬时的

cumulative［'kjuːmjulətiv］adj. 累积的

circumferential［səˌkʌmfə'renʃəl］adj. 圆周的

curvilinear［ˌkəːvi'liniə(r)］adj. 曲线的

turnbuckle［'təːnˌbʌkl］n. 螺丝扣，套筒螺母

stave［'steiv］n. 窄板，桶材

Notes：

① reinforcing bars，steel bars，reinforcement，reinforcing steel 钢筋

② cast-in-place concrete 现浇混凝土

③ precast concrete 预制混凝土

④ a plain concrete beam 素混凝土梁

⑤ a reinforced concrete member 钢筋混凝土构件

⑥ slenderness 长细比

⑦ post-tensioning method 后张法
⑧ pre-tensioning method 前张法

Exercises:

1. Translate the Following Sentences into Chinese.

（1）As a result, cracks develop whenever loads, or restrained shrinkage or temperature changes, give rise to tensile stresses in excess of the tensile strength of the concrete.

（2）The form must be strong enough to support both the weight and hydrostatic pressure of the wet concrete, and any forces applied to it by workers, concrete buggies, wind and so on.

（3）Frequently, however, the overall cost is affected as much or more by the overall construction time since the contractor and owner must borrow or otherwise allocate money to carry out the construction and will not receive a return on this investment until the building is ready for occupancy.

（4）Because there is less mass of steel to be heated or cooled, and because steel is a better conductor than concrete, a steel structure is generally affected by temperature changes to a greater extent than is a concrete structure. On the other hand, concrete undergoes drying shrinkage, which, if restrained, may cause deflections or cracking.

（5）Unless care is taken in design and construction, however, these cracks may be unsightly or may allow penetration of water. When this occurs, water or chemicals such as road deicing salts may cause deterioration or staining of the concrete.

（6）The sections are then able to behave elastically, and almost the full capacity of the concrete in compression can be efficiently utilized across the entire depth of the concrete sections when all loads act on the structure.

（7）Such an imposed longitudinal force is called a prestressing force, i.e., a compressive force that prestresses the sections along the span of the structural element prior to the application of the transverse gravity dead and live loads or transient horizontal live loads.

（8）The reinforcement in the reinforced concrete member does not exert any force of its own on the member, contrary to the action of prestressing steel.

（9）Structures designed to withstand heavy vibrations, such as machine foundations, can easily be made rigid through the contribution of the prestressing force to the reduction of their otherwise flexible deformation behavior.

（10）In spite of these additional costs, if a large enough number of precast units are manufactured, the difference between at least the initial costs of prestressed and reinforced concrete systems is usually not very large.

（11）He subsequently developed the idea that successive post-tensioning of unbonded rods would compensate for the time-dependent loss of stress in the rods due to the decrease in the length of the member because of creep and shrinkage.

2. Translate the Following Sentences into English.

（1）混凝土的抗拉强度远低于其抗压强度，因此混凝土很容易产生开裂。

（2）在钢筋混凝土梁中，钢筋埋置于混凝土中，使当混凝土开裂后平衡弯矩所需的拉力由钢筋承担。

（3）由于钢筋放入模板后才浇筑混凝土，因此混凝土必须能够流过钢筋及模板并完全充满模板的每个角落。

（4）在相同的跨度和荷载条件下，预应力构件的截面高度比钢筋混凝土构件要小。

15　Concrete: Forming, Reinforcing, Placement and Curing
（混凝土：支模、扎筋、浇捣和养护）

教学目标：本单元的主要内容有混凝土支模、扎筋、浇捣和养护。教学重点是混凝土养护，难点是混凝土的配合比设计。通过本单元的学习，要求掌握混凝土支模、扎筋、浇捣和养护的整个施工过程，特别要注意混凝土养护是混凝土中水泥的水化过程，了解混凝土的配合比设计。

15.1　Forming

The forming cost component of concrete work can be significant. For relatively complicated concrete work, labor and material forming costs may be equal to as much as 70% of the total cost. Concrete forming is characterized by the numerous alternative forming systems available to a contractor to perform a specific type of work. There may be literally over 100 alternative forming systems available to a contractor to form a defined concrete member. Perhaps there is no other area of construction work that offers the contractor more flexibility in regard to a work method.

Because of this, one of the more difficult estimating problems the estimator faces is selection of the most effective forming system for the work in question. A related problem is the need for the estimator to be correct in his assumption that the forms he estimates will be the forms the superintendent will in fact use. If this is not true, there is a good possibility that the cost of forms put in place will vary significantly from the cost of those estimated. Concrete forming differs from other types of construction work because the put-in-place forms are temporary. Unlike the other types of construction work that is placed permanently in the structure, concrete forms usually are erected and subsequently taken apart (this is referred to as the stripping of the forms). The temporary characteristic of forming creates two somewhat unique estimating problems.

First, the estimator must be careful to include in his labor cost the erection, stripping and any related cleaning and/or storing costs. The labor time required to strip, clean and store forms may approach the labor time required to erect the forms.

Second, a more difficult aspect of estimating formwork—a problem that relates to the temporary characteristic of the in-place forms—is the fact that forms are often re-used several times for a single project. It is common for the constructor to re-use a single form four or more times on a single project. This means that the estimator must be able to estimate the number of actual re-used of concrete forms on a project as well as the material cost of the forms and the labor cost of erecting them. A form that is re-used twice on a project in effect has a material cost twice that if the form are re-used 4 times. That is, the material cost per use is dependent on the cost of the form and the number of times the form is used.

Formwork is most commonly measured or cost in units of square meter of contact area. A concrete wall that is 2.5m high and 3.0m long that needs to be formed on both sides will require $15m^2$ of contact area. Concrete forming is characterized by many alternative systems that are available to the builder. While many forming systems are available for different types of members (for example, walls, beams, columns and slabs), the various systems are often characterized as being either job-built or pre-built. A job-built forming system is essentially custom-designed at its placement of location for the concrete. Plywood or sheathing, boards and nails are fabricated at the job site in a job-built forming system.

Job-built forming systems are commonly constructed with sheathing and plywood. Given the flexibility to cut the materials, job-built systems can be custom-designed to almost any shape or design. These systems require a significant amount of labor time and cost to fabricate. The labor cost for building the forms is typically greater than the material cost for the materials. Job-built forms made from sheathing or plywood are commonly used 4 to 8 times on a single project. Job-built forms are used to form walls, beams, columns, slabs or any other member.

Contrasted to job-built forms are pre-built or pre-fabricated forms. Unlike many job-built forms, a contractor typically re-uses a pre-built form many times; much as a truck or a crane. In other words, a pre-built form might be considered as equipment by the contractor rather than material. Pre-built forms are often made of a combination of steel or a related metal and a wood product, including plywood. Numerous manufacturers have patented similar pre-built systems; many of the systems compete with one another. Pre-built forms made from a combination of steel and wood are sometimes referred to as *a steel-ply system.*[①] While steel-ply pre-built forms are predominant for wall systems, fiberglass and paper products are often used for pre-built beam and column systems. Metal pan systems are extensively used for pre-built slab forms.

Unlike job-built forms, the relatively high cost of a pre-built system often results in a higher material cost for pre-built forms than the labor cost for erecting the form. The reuse factor and perhaps a lower erection cost are the reasons why a pre-built form may cost less than a job-built form for a specific application. The economics of job-built versus pre-built forming systems is dependent on many factors, including the amount and type of forming to be done. By now it should be apparent that there is the potential for an estimator to define numerous separate work items when taking off quantities and pricing them. However, given estimating time constraints, too long a list is not feasible. In defining concrete forming work items, the estimator should consider the following factors in delineating separate work items: (1) type of member or structure being formed; (2) specific type of form that is to be used (i.e., pre-built versus job-built, etc.); (3) characteristics or location of the forming that causes it to be unique.

It is true that some other considerations also likely affect the price of forming and thus might be considered in delineating specific work items. For example, the quality of the concrete finish that is to be obtained for a wall surface might be considered in estimating formwork. However, this and other factors can be recognized in the pricing of specific work items rather than as a basis for work-item definition.

15. 2　Reinforcement

The majority of concrete used in building construction is designed to be a structural bearing material. Given the fact that concrete is relatively weak in regard to resisting tension load, most concrete applications include a form of structural reinforcing, such as reinforcing bars, wire mesh or steel wire tendons. The use of tendons is usually limited to *pre-stressed concrete.*[②] The estimating of tendons usually causes the estimator few problems because pre-stressed concrete is often purchased as a completed unit. Most construction projects call for reinforced concrete. Reinforcing, like concrete forming, is done before the concrete is placed. However, unlike concrete forming, the steel reinforcement is left in the concrete.

Steel reinforcement embedded in concrete is usually designed to carry the tension or bending loads imposed on the finished concrete member. Reinforcement sizes have been standardized and are usually quoted in the diameter of the bar. For example, the diameter of a $\phi6$ bar is 6mm. Owing to the uniqueness of designed concrete members, there are no standardized amounts or sizes of reinforcement to be placed in concrete member. The result is that the ·estimator cannot merely take off the quantity of concrete to be placed and use a factor to determine the reinforcement to be purchased and fabricated. Instead, more often than not, the estimator has to take off the reinforcements separately from his determination of the formwork or concrete quantity calculation.

The reinforcement work to be performed is designed on the construction drawings by the designer; it specifies the length of reinforcement to be placed and the size of reinforcement (e.g., $\phi5$ bar), along with its location. While designated via a listing of its diameter, reinforcement is purchased and priced via a weight calcula-

tion; the contractor usually purchases reinforcement at a price per ton. This price per ton may also reflect the amount of factory bends that are to be made, the distance the reinforcement has to be shipped, and the amount of variation in the number of sizes ordered. However, the total weight of the reinforcement is the predominant factor in determining the purchase price. There is a need to convert a reinforcement take-off from the length factor into a weight calculation. In this regard, various reinforcement sizes are combined to yield a total weight of reinforcing required. In order to convert reinforcing sizes and lengths into a weight calculation, it is necessary for the estimator to use a conversion factor.

15.3　Placement

When we think of concrete construction work, we often focus on its actual placement, whether this entails *pumping the concrete*[3], placing it with the use of a crane bucket, or using an alternative production process. However, as has been pointed out in this chapter, the placement of concrete is only one of three processes that are typical of concrete construction. In fact, in regard to the labor cost component of the work, the placement of the concrete is typically less time-consuming and expensive than the forming or placement of reinforcement. Nonetheless, concrete placement remains an important work process in regard to the need to accurately estimate its cost. If it were not for the need to place concrete there would be no forming or placement of reinforcement work processes.

As noted earlier, concrete is fabricated for a building either in the precast or the cast-in-place state. *Pre-cast concrete*[4] is essentially a totally prefabricated unit that includes the embedded reinforcing. The estimating of pre-cast concrete causes few difficulties. The quantity of work is usually easily determined via the project

drawings, and the purchase price is usually well defined because the pre-cast members are commonly purchased from an outside vendor. Therefore, perhaps the only variable at issue is the labor time and cost required to erect the pre-cast members. Given the specialty nature of the work, pre-cast concrete work should be delineated as a separate work item from cast-in-place concrete work. Because of its somewhat infrequent usage, it may be sufficient to identify a single work item for pre-cast concrete.

Owing to its more common usage and the relatively more complicated estimating process, the estimator likely needs to be more attentive to the estimating of cast-in-place concrete than to that of pre-cast concrete. Cast-in-place concrete is essentially concrete that is poured in a liquid state at the job site and finished and cured into a hardened state. Concerning the estimating of cast-in-place concrete, two types of estimates may be required. For one, it might be the contractor's responsibility or decision to produce the concrete himself; that is, purchasing the concrete ingredients of sand, water and cement, and mixing them to yield the concrete to be poured. The proportioning of and mixing of the ingredients of concrete to yield finished material is commonly referred to as *mix design*. [5] Often the mix-design process is outside the responsibility of the building contractor. Instead, as the second alternative (which also affects the estimate), the contractor purchases the concrete already mixed from a manufacturer of the concrete. This is referred to as *ready-mix concrete*. [6]

A construction project has to be of significant size to justify the contractor's setting up a *concrete batching plant* [7] that is capable of proportioning and mixing concrete at the job site. While this practice may be somewhat common on heavy and highway projects, it is rare for a building project. When it is done, the estimator must in fact understand mix design in order to accurately estimate both the material and labor cost of producing the concrete. Because this process is rare in the building process, following paragraphs of this article will assume that the contractor will purchase ready-mix concrete for the cast-in-place process.

Even using ready-mix concrete, the estimator is still faced with the preparation of an estimate for cast-in-place concrete. But because the concrete is purchased from an outside vendor, estimating the concrete material cost presents few problems. Often a contractor enters into a contract with a ready-mix producer for a fixed price per cubic yard of concrete before he starts work on a project, or even before the submission of a bid for the project in question. Thus, perhaps the only difficulty the estimator has in determining the material cost is to estimate the quantity from the construction drawings and then to determine the material wastage factor.

Both the material cost and the labor cost for placing concrete are related to the volume of concrete to be placed. It is for this reason that cast-in-place concrete is usually taken off in cubic yards of material required. The labor cost required to place the concrete is more difficult to estimate. It is dependent upon the factors that affect any type of construction work, including labor morale, degree of supervision, weather conditions and so on. In addition to being subject to these conditions, the determination of the labor cost for placing concrete is made more complex by the fact that there are numerous different methods of placing concrete. For example, a crane, pump or perhaps a conveyor might all be used to place a specific concrete pour.

The contractor is almost always given the freedom to use whatever method, equipment and labor crew he wants in regard to making a concrete pour. Obviously, his choice is somewhat constrained by the availability of equipment or labor resources. However, given the availability of rental equipment, the contractor usually has several alternative feasible methods. Even if he limits his choice to a specific type of construction method—say, a crane—numerous types of cranes with varying production capacities are likely to be available. Each of these and accompanying required labor crews involve different costs. The use of alternative means of placing concrete presents the estimator with several problems. For one, the estimator has to be able to know the production rates for the feasible alternative methods.

A second problem relates to the possible lack of communication that sometimes exists between a contractor's estimating personnel and project production personnel. There is a possibility that the estimator might assume the use of one construction method in preparing his estimate, only to find later that the project superintendent has chosen a different method. For example, the estimator might assume the use of a crane in preparing his estimate, whereas the superintendent in fact plans to use a pump to place the concrete. Given different costs for the two methods, the result is that the estimate is incorrect in regard to the cost of performing the work. It should be obvious that there is a strong need for communication in regard to the estimating and production functions. Naturally, this is true in regard to all the construction work that is to be estimated and built. However, with the numerous possible alternatives for a specific type of concrete placement, the need for the communication is emphasized for the type of work.

15.4 Curing

In order to obtain good concrete the placing of an appropriate mix must be followed by curing in a suitable environment during the early stages of hardening. Curing is the name given to procedures used for promoting the hydration of cement. It consists of controlling temperature and moisture movement into and out of the concrete. *More specifically, the object of curing is to keep concrete saturated, or as nearly saturated as possible, until the originally water-filled space in the fresh cement paste has been filled to the desired extent by the products of the hydration of the cement.* [8] In the case of site concrete, active curing nearly always stops long before the maximum possible hydration has taken place.

The necessity for curing arises from the fact that the hydration of cement can take place only in water-filled capillaries. For this reason, a loss of water by evaporation from the capillaries must be prevented. Furthermore, water lost internally by self-desiccation has to be replaced by water from the outside, i.e., the ingress of water into the concrete must be made possible.

It may be recalled that the hydration of a sealed specimen can proceed only if the amount of water present in the paste is at least twice that of the water already combined. *Self-desiccation is thus of importance in mixes with water/cement ratios below about 0.5; for higher water/cement ratios the rate of hydration of a sealed specimen equals that of a saturated specimen.* [9] It should not be forgotten, however, that only half the water present in the paste can be used for chemical combination; this is so, even if the total amount of water present is less than the water required for chemical combination. *This statement is of considerable importance as it was formerly thought that, provided a concrete mix contained water in excess of that required for the chemical reactions with cement, a small loss of water during hardening would not adversely affect the process of hardening and the gain of strength.* [10]

Once the concrete has set, wet curing can be provided by keeping the concrete in contact with a source of water. This may be achieved by spraying or flooding (ponding), or by covering the concrete with wet sand or earth, sawdust or straw. Periodically-wetted hessian or cotton mats may be used, or alternatively an absorbent covering with access to water may be placed over the concrete. A continuous supply of water is naturally more efficient than an intermittent one.

Another means of curing is to use an impermeable membrane or waterproof paper. A membrane, provided it is not punctured or damaged, will effectively prevent evaporation of water from the concrete but will not allow the ingress of water to replenish that lost by self-desiccation. The membrane is formed by sealing compounds which may be clear, white or black. The opaque compounds have the effect of shading the concrete, and a light color leads to a lower absorption of the heat from the sun, and hence to a smaller rise in the temperature of the concrete. The period of curing cannot be prescribed simply, but it is usual to specify a minimum of seven days for ordinary Portland cement concrete. With slower-hardening cements a longer curing period is desirable. High-strength concrete should be cured at an early age as partial hydration may make the capillaries discontinuous: on the renewal of curing, water would not be able to enter the interior of the concrete and no further hydration would result. However, mixes with a high water/cement ratio always retain a large volume of capillaries so that curing can be effectively resumed at any time. No loss of strength is caused by delaying the curing.

Vocabulary and Expressions：

cast-in-place 现浇

cure [kjuə] v. 养护

custom designed 定制

delineate [di'linieit] vt. 描绘

discretion [dis'kreʃən] n. 斟酌，判断，慎重

forming ['fɔ：miŋ] n. 成型，支模板；~ system 模板体系；stripping of the ~ 拆模

formwork ['fɔ：mwə：k] n. 模板工程

in-place 现场

interpolate [in'tə：pəleit] vt. 内插

job-built 现场制作

placement ['pleismənt] n. 浇捣

plywood ['plaiwud] n. 胶合板

pre-built 预制

sheathing ['ʃi：ðiŋ] n. 护套

superintendent [ˌsju：pərin'tendənt] n. 管理员，指挥人，总段长

hardening ['hɑ：dniŋ] n. 硬化

hydration [hai'dreiʃn] n. 水化，水化作用

membrane ['membrein] n. 膜，薄膜，羊皮纸

desiccation [ˌdesi'keiʃn] n. 干燥，干缩，干裂

hessian ['hesiən] n. 粗麻布，浸沥青的麻绳

capillary [kə'piləri] n. 毛细管，毛细管现象

Notes：

① a steel-ply system 折叠式钢模板系统

② pre-stressed concrete 预应力混凝土

③ pumping the concrete 用泵输送混凝土

④ pre-cast concrete 预制混凝土

⑤ mix design 配合比设计

⑥ ready-mix concrete 商品(或预拌)混凝土

⑦ concrete batching plant 混凝土搅拌站

⑧ 更确切地说，养护的目的是要使混凝土保持或尽可能接近于饱和状态，直至新鲜水泥浆中原始充水空间为水泥水化物填充到所要求的程度为止。

⑨ 因此，自干作用在水灰比低于 0.5 左右

的拌合物中才有其重要性；至于在高水灰比的情况下，密封试件的水化速度则与饱和试件的水化速度相同。

⑩ 这一说明十分重要，因为原先曾认为只要混凝土拌合物的含水量超过与水泥水化反应所需用水量，则在硬化期间即使有少量的失水也不会对硬化过程和强度的增长产生有害的影响。

Exercises：

1. Translate the Following Sentences into Chinese.

(1) The hardness, strength and binding properties of cement are due to the formation of certain compounds which are produced as a result of chemical action between cement and water.

(2) To prevent evaporation, either the surface is coated with a layer of bitumen or similar other waterproof compound which fills the concrete pores and does not allow the water to evaporate; or the surface is kept covered with waterproof paper or membrane or with a layer of wet sand, wet clay or gunny bags.

(3) It must be clearly understood that the absence of proper curing of cement concrete can greatly undermine its ultimate strength and that many failures have occurred in the past on this account.

(4) From the point of view of curing, it is advantageous to keep the forms on for as long a period as possible as they help in retaining moisture.

2. Translate the Following Sentences into English.

(1) 模板工程绝大多数情况下是以接触面的平方米为单位来计量和计费的。

(2) 像模板工程一样，钢筋工程是在浇筑混凝土前进行的，然而与模板工程不同的是，钢筋是留在混凝土中的。

(3) 在浇筑混凝土前，通常需要进行混凝土的配合比设计，当然也可以购买商品混凝土，而且这已成为一种趋势。

16　Structural Systems of Buildings

（建筑结构体系）

教学目标： 本单元的主要内容有双向板、建筑结构体系、框架、剪力墙、刚架、框剪、框筒、外部斜撑筒、成束筒、桁架、拱、折板、壳以及膜结构。教学重点是建筑结构体系，难点是膜结构。通过本单元的学习，要求熟悉框架、剪力墙、刚架、框剪、框筒、外部斜撑筒、成束筒、桁架、拱、折板、壳以及膜等各种建筑结构，了解建筑结构体系。

The structure of a building can be defined as the assemblage of those parts which exist for the purpose of maintaining shape and stability. Its primary purpose is to resist any loads applied to the building and to transmit those to the ground.

The structure must also be engineered to maintain the architectural form. There are at least three items that must be present in the structure of a building: stability, strength and stiffness and economy. Apart from these three primary requirements, several other factors are worthy of emphasis. First, the structure or structural system must relate to the building's function. It should not be in conflict in terms of form. Second, the structure must be fire-resistant. Third, the structure should integrate well with the building's circulation systems. It should not be in conflict with the piping systems for water and waste, the ducting systems for air or (most important) the movement of people. It is obvious that the various building systems must be coordinated as the design progresses. Fourth, the structure must be psychologically safe as well as physically safe.

Analogous to steel or composite construction, concrete offers a wide range of structural systems suitable for high-rise buildings. There are perhaps as many structural concepts as there are engineers, making it awkward if not impossible to classify all the concepts into distinct categories. However, for purposes of presentation, it is convenient to group the most common systems into separate categories, each with an applicable height range as shown in Fig. 16-1. Although the height range for each group is logical for normally proportioned buildings, the appropriateness of each system to a particular building can only be judged when all other factors influencing the lateral load behavior are taken into account. *Such factors include building geometry, severity of exposure to wind, seismicity of the region, ductility of the frame, and limits imposed on the size of the structural members.* [1] Oft-times, systems combining the characteristics of two or more can be employed to fulfill the specific project requirement. The multitude of the systems available presents an opportunity for an experienced engineer to come up with a structural system that will serve its optimum function in the overall sense on the project. Although the selection of a system requires knowledge of both horizontal and lateral systems, the material presented in this paper emphasizes the requirements of lateral systems only.

Fig. 16-1 shows 14 different categories of structural systems, starting with the most elementary system consisting of floor slabs and columns. *At the end of the spectrum is the bundled tube system, which is appropriate for very tall buildings and for buildings with a large plan aspect ratio.* [2]

16.1 Frame Action of Column and Two-Way Slab Systems

Concrete floors in tall buildings often consist of a two-way floor system such as a flat plate, flat slab or a waffle system. In a flat plate system the floor consists of a concrete slab system of uniform thickness which is framed directly into columns. The flat slab system makes use of either column capitals, drop panels or both to increase the shear and moment resistance of the system at the columns where the shears and moments are greatest. The waffle slab consists of two rows of joists at right angles to each other commonly formed by using square domes. The

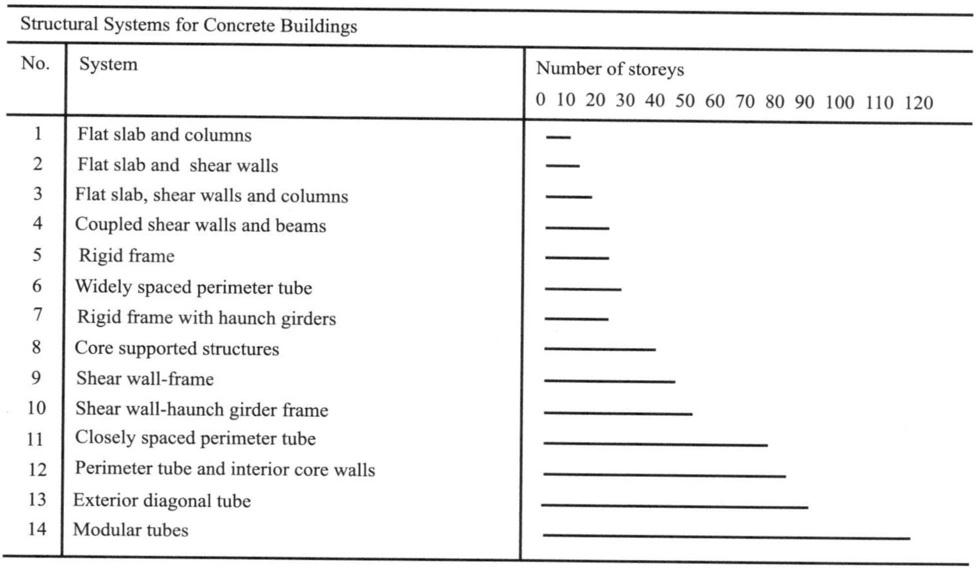

Structural Systems for Concrete Buildings		
No.	System	Number of storeys 0 10 20 30 40 50 60 70 80 90 100 110 120
1	Flat slab and columns	▬
2	Flat slab and shear walls	▬
3	Flat slab, shear walls and columns	▬▬
4	Coupled shear walls and beams	▬▬
5	Rigid frame	▬▬
6	Widely spaced perimeter tube	▬▬▬
7	Rigid frame with haunch girders	▬▬
8	Core supported structures	▬▬▬
9	Shear wall-frame	▬▬▬▬
10	Shear wall-haunch girder frame	▬▬▬▬
11	Closely spaced perimeter tube	▬▬▬▬▬
12	Perimeter tube and interior core walls	▬▬▬▬▬▬
13	Exterior diagonal tube	▬▬▬▬▬▬
14	Modular tubes	▬▬▬▬▬▬▬

Fig. 16-1   Structural Systems for Concrete Buildings

domes are omitted around the columns to increase the moment and shear capacity of the slab. Any of the three systems can be used to function as an integral part of the wind-resisting systems for buildings in the 10-storey range.

The slab system shown in Fig. 16-2 has two distinct actions in resisting lateral loads. First, because of its high in-plane stiffness, it distributes lateral loads to various vertical elements in proportion to their bending stiffness. Second, because of its significant out-of-plane stiffness, it restrains the vertical displacements and rotations of the columns as if they are interconnected by a shallow wide beam.

The concept of " effective width " as explained below can be used to determine the equivalent width of the slab. Although physically no beam exists between the columns, for analytical purposes it is convenient to consider a certain width of slab as a beam framing between the columns. The effective width is however, dependent on various parameters, such as column aspect ratios, distance between the columns, thickness of the slab, etc. Research has shown that values less than, equal to and greater than full width are all valid depending upon the parameters mentioned above.

Note that the American Concrete Institute (ACI) code permits the full width of slab between adjacent panel center lines for both gravity and lateral loads. The only stipulation is that the two-way systems analysis should take into account the effect of slab cracking in evaluating the stiffness of frame members. Use of full width is explicit for gravity analysis and implicit for the lateral loads.

However, engineers generally agree that using full width of the slab gives unconservative results for lateral load analysis. The method tends to overestimate the slab stiffness and underestimate the column stiffness, producing the error in estimating the distribution of moments due to lateral loads.

The shortcomings of using full width in lateral load analysis can be overcome by determining the equivalent stiffness on the basis of a finite-element analysis. The stiffness thus obtained is appropriate for both gravity and lateral analysis.

Of particular concern in the detailing of two-way systems is the problem of stress concentration at the slab-column joint where, especially under the vertical load, nonlinear behavior is initiated concrete cracking and steel yielding. Shear reinforcement at the column-slab joint is

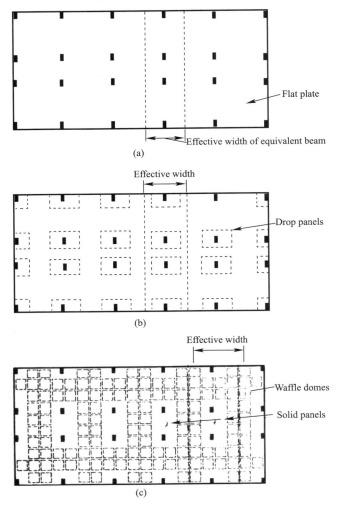

Fig. 16-2　Lateral Systems Using Slab and Columns
(a) Flat Plate; (b) Flat Slab with Drop Panels; (c) Two-Way Waffle System

necessary to improve the joint behavior and avoid early stiffness deterioration under lateral cyclic loading. Note that two-way slab systems without beams are not permitted by ACI code in regions of high seismic risk (zones 3 and 4). Their use in regions of moderate seismic risk (zone 2) is permitted subject to certain requirements, mainly relating to reinforcement placement in the column strip. Since the requirements are too detailed to be described here the reader is referred to the UBC and ACI codes for further details.

16.2　Structural Systems of Buildings

(1) Frames: A frame structure is one made up of linear elements, typically beams and columns, which are connected to one another at their ends, called joints, which do not allow any relative rotations between the ends of the attached members, although the joints themselves

may rotate as a unit. Frames are applicable to both large and small buildings. Fig. 16-3 shows some types of frames. One of the simplest types of frames is a single span frame consisting of two columns and a rigidly connected beam. The frame with a beam that is constructed as two columns and a peaked roof is termed portal or gable frame. The concept of the single span frame may be extended to a multi-cell frame. For example, it can be widened horizontally to form a multi-bay frame and/or raised vertically to form a multi-storey frame.

A frame structure is able to resist horizontal loads as well as vertical loads. When the beam is subjected to a vertical load, the load enables it to be flexural and tends to cause its end to rotate. The column tops and beam ends

are however connected rigidly. Free rotation at the end of the beam cannot take place because the columns tend to prevent or restrain the beam end from rotating. As a result, the columns must carry bending moments as well as axial forces, and then transfer them to the ground. When a frame structure is subjected lateral loads, the beam restrain the columns from freely rotating in a way that would lead to the total collapse of the structure by means of the presence of rigid connections. The stiffness of the beam has much to do with the lateral-load-bearing resistance of a frame. It also serves to transfer part of the lateral load from one to the other. The action of lateral loads produces bending, shear and axial forces in all members of a frame.

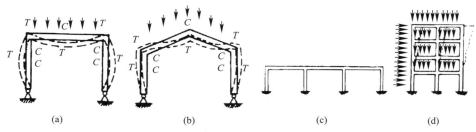

Fig. 16-3   Types of Frames
(a) Frame; (b) Portal Frame; (c) Multi-Bay Frame; (d) Multi-Storey Frame

（2）Shear walls：The applicable height range of slab and shear wall systems can be increased marginally by including the frame action of column and slabs. The system is best suited for apartments, condominiums and hotels, and is identical to the system described in the previous section. The difference is only in the analysis in that the frame action of column and slabs is also taken into account in the lateral load analysis. Whether this action is significant or not is a function of relative stiffness of various elements. In most apartment or hotel layouts, the frame resistance to overturning moments is no more than 10% to 20% of the resistance offered by shear walls. Many engineers, therefore, ignore the frame action altogether by designing the shear walls to carry the total lateral loads. However, in keeping with the current trend of taking

advantage of all available structural actions, it is advisable to include the frame action in analysis.

When two or more shear walls are interconnected by a system of beams or slabs, the total stiffness of the system exceeds the summation of the individual wall stiffness, because the connecting slab or beam restrain the individual cantilever action by forcing the system to work as composed unit (Fig. 16-4). Such an interacting shear wall system can be used economically to resist lateral loads in buildings up to about 40 storeys. However, planar shear walls are efficient lateral load carriers only in their plane. Therefore, it is necessary to provide walls in two orthogonal directions. However, in long and narrow buildings sometimes it may be possible to resist wind loads in the long direction by the frame action of columns and slabs because first,

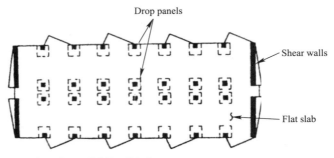

Fig. 16-4　Shear Wall Flat Slab System

the area of the building exposed to the wind is small, and second, the number of columns available for frame action in this direction is usually large. The layout of walls and columns should take into consideration the torsional effects.

Walls around elevators, stairs and utility shafts offer an excellent means of resisting both lateral and gravity loads without requiring undue compromises in the leasability of buildings. Closed- and partially closed-section shear walls are efficient to resisting torsion, bending moments and shear forces in all directions, especially when sufficient strength and stiffness are provided around door openings and other penetrations.

(3) Rigid frame: Cast-in-place concrete buildings have the inherent advantage of continuity at joints. Girders framing directly into columns, can be considered rigid with the columns; such a girder-column arrangement can be thought of as a portal frame. However, girders that carry shear and bending moments due to lateral loads often require additional construction depth, necessitating increases in overall height of the buildings.

The design and detailing of joints where girders frame into building columns should be given particular attention, especially when buildings are designed to resist seismic forces. The column region within the depth of the girder is subjected to large shear forces. Horizontal ties must be included to avoid uncontrolled diagonal cracking and disintegration of concrete. Specific detailing provisions are given in the UBC and ACI codes to promote ductile behavior

in high-risk seismic zones 3 and 4, and somewhat less stringent requirements in moderate-risk seismic zone 2. The underlying philosophy is to design a system that can respond to overloads without loss in gravity-load carrying capacity.

Rigid frame system for resisting lateral and vertical loads have long been accepted as a standard means of designing buildings because they make use of the stiffness in the beams and columns that are required in any case to carry the gravity loads. In general, rigid frames are not as stiff as shear wall construction and are considered more ductile and less susceptible to catastrophic earthquake failures when compared to shear wall structure.

A rigid frame is characterized by its flexibility due to flexure of individual beams, columns and rotation at their joints. The strength and stiffness of the frame is proportional to the beam and column size and inversely proportional to the column spacing. Internally located frames are not very popular in tall buildings because the leasing requirements of most buildings limit the number of interior columns available for frame action. The floor beams are generally of long spans and are of limited depth. However, frames located at the building exterior do not necessarily have these disadvantages. An efficient frame action can thus be developed by providing spaced columns and deep beams at the building exterior.

(4) Shear wall—frame interaction: Without question this system is one of the most, if not the most, popular system for resisting lateral loads. The system has a broad range of applica-

tion and has been used for buildings as low as 10-storey to as high as 50-storey or even taller buildings. With the advent of haunch girders, the applicability of the system is easily extended to buildings in 70- to 80-storey range.

　　The interaction of frame and shear walls has been understood for quite some time; the classical mode of the interaction between a prismatic shear wall and a moment frame is shown in Fig. 16-5; the frame basically deflects in a so-called shear mode while the shear wall predominantly responds by bending as a cantilever. Compatibility of horizontal deflection produces interaction between the two. The linear sway of the moment frame, when combined with the parabolic sway of the shear wall results in an

enhanced stiffness because the wall is restrained by the frame at the upper levels while at the low levels the shear wall is restrained by the frame. However, it is not always easy to differentiate between the two modes because the frame consisting of closely spaced columns and deep beams tends to behave more like a shear wall responding predominantly in a bending mode. And similarly, a shear wall weakened by large openings may tend to act more like a frame by deflecting in a shear mode. The combined structural action, therefore, depends on the relative rigidity of the two, and their modes of deformation. Furthermore, the simple interaction diagram given in Fig. 16-5 is valid only if:

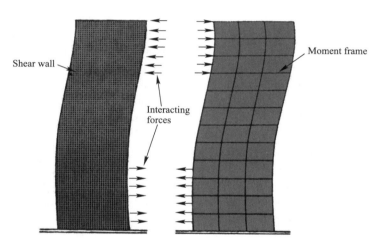

Fig. 16-5　Shear Wall-Frame Interaction

　　1) the shear wall and frame have constant stiffness throughout the height;

　　2) if stiffness varies, the relative stiffness of the wall and frame remains unchanged throughout the height.

　　Since architectural and other functional requirements frequently influence the configuration of structural elements, the above conditions are rarely met in a practical building. In a contemporary high-rise building, very rarely the geometry of walls and frames be the same over the full height. For example, walls around the elevators are routinely stopped at levels corresponding to the elevator's drop, columns are

made smaller as they go up, and the building geometry is very rarely the same for the full height. Because the abrupt changes in the stiffness of walls and frames combined with the variations in the geometry of the building, the simple interaction shown in Fig. 16-5 does not even close to predicting the actual behavior of the building structures. However, with the availability of two- and three-dimensional computer programs, capturing the essential behavior of the shear wall-frame system is within the reach of everyday engineering practice.

　　(5) Frame tube structures: The tube concept is an efficient framing for taller slender

buildings. In this system, the perimeter of the building consisting of closely spaced columns works as a relatively deep spandrel. *The resulting system works as a giant vertical cantilever and is very efficient because of the large separation between the windward and leeward columns.* [③] The tube concept in itself does not guarantee that the system satisfied stiffness and vibration limitations. The "chord" drift caused by the axial displacement of the columns and the "web" drift brought about by the shear and bending deformations of the spandrels and columns may vary considerably depending upon the geometric and elastic properties of the tube. *For example, if the plan aspect ratio is large, say much in excess of 1 : 1.5, it is likely that a supplemental lateral bracing is necessary to satisfy drift limitations.* [④] The number of storeys that can be achieved economically by using the tube system depends on a number of factors such as spacing and size of columns, depth of perimeter spandrels and plan aspect ratio of the building. The system should be given serious consideration for buildings taller than about 40-storey.

(6) Exterior diagonal tube: Master builder Fazlur Khan of Skidmore, Owings & Merrill envisioned as early as 1972 that it was possible to build high rises in concrete rivaling those in structural steel. His quest to find a structural solution for eliminating the shear lag phenomenon led him to the diagonal tube concept. A brilliant manifestation of this principle in steel construction is seen in the John Hancock Tower in Chicago. Applying similar principles, Khan visualized a concrete version of the diagonal truss tube consisting of exterior columns spaced at about 10ft (3.05m) centers with blocked out windows at each floor to create a diagonal pattern on the facade. The diagonals could be designed to carry the shear forces, thus eliminating bending in the tube columns and girders. Although Khan enunciated the principle in the 1970s, the idea had to wait 13 years to find its way to a real building. Currently, two high rises have been built using this approach. The first is a 50-storey office structure located on Third Avenue in New York and the second is a mixed-use building located on Michigan Avenue in Chicago. The structural system for the building in New York consists of a combination of framed and trussed tube interacting with a system of interior core walls. All the three subsystems, namely, the framed tube, trussed tube and shear walls, are designed to carry both lateral and vertical loads. *The building is 570ft (173.73m) high with an unusually high height-to-width ratio of 8 : 1.* [⑤] The diagonals created by filling in the windows serve a dual function. First, they increase the efficiency of the tube by diminishing the shear lag, and second, they reduce the differential shortening of exterior columns by redistributing the gravity loads. A stiffer, much more efficient structure is realized with the addition of diagonals. The idea of diagonally bracing this structure is suggested by Fazlur Khan to the firm of Robert Rosenwasser Associates, who executes the structural design for the building. Schematic elevation of the building is shown in Fig. 16-6.

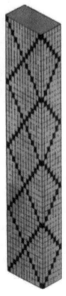

Fig. 16-6　Schematic Elevation of Exterior Braced Tube

The Chicago version of the braced concrete tube is a 60-storey multi-use project. The building rises in two tubular segments above a flared base. According to the designers, diagonal bracing is used primarily to allow maximum flexibility in the interior layout needed for mixed uses. In contrast to the building in New York, which has polished granite as cladding, the Chicago building sports exposes concrete framing and bracing.

(7) Modular or bundled tube: The concept of bundled tube in concrete high rises is to connect two or more individual tubes in a bundle with the object of decreasing the shear lag effects. Fig. 16-7(a) shows a schematic plan of a bundled tube structures. Two versions are possible using either framed or diagonally braced tubes as shown in Fig. 16-7(b) and (c). A mixture of the two is, of course, possible.

(8) Trusses: A truss is a structure made by assembling individual linear elements arranged in a triangle or combination of triangles to form a framework. The elements are typically assumed to join at their intersections with pinned connections. There are many possible configurations of trusses as shown in Fig. 16-8. Both the top and bottom members are termed chords and the members between chords are

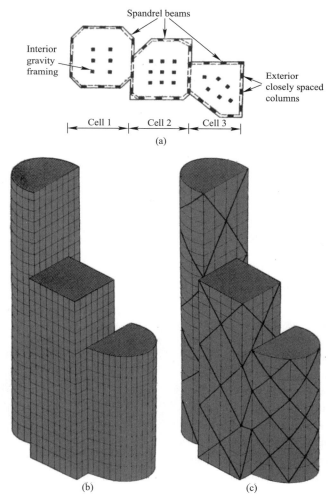

Fig. 16-7　Bundled Tube
(a) Schematic Plan; (b) Framed Bundled Tube; (c) Diagonally Braced Bundled Tube

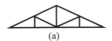 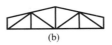 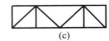

Fig. 16-8　Types of Plane Trusses
(a) Triangular Truss; (b) Trapezoidal Truss; (c) Parallel-Chord Truss

termed webs. The primary principle of using the truss is that arranging members into some triangular configurations results in a stable structure. A truss made up of members is bent in an overall way under an applied loading which is much the same way as a beam bending. However, members are not subjected to bending but are either purely in compression or tension.

Trusses can also be in the form of spatial structures to carry loads. A space truss is usually a large span of surface structure that is composed of rigid members triangulated in three dimensions and arranged in repeating geometric units. There are many ways of arranging repetitive geometric units to form different types of space trusses. Fig. 16-9 shows one type of space truss.

Fig. 16-9　Space Truss

(9) Cables: Cables are flexible line-forming structural elements. A cable subjected to external loads will obviously deform in a way dependent on the magnitude and location of the loads. The form obtained is often termed the funicular shape of the cable. The term funicular is derived from the Latin word "rope". Only tension forces are developed in the cable. When a cable is used to span two points and carry an external concentrated load or more, it deforms into a shape made up of a series of straight-line segments. A cable with constant cross section carrying only its own self-weight naturally deforms into a catenary shape, whereas a cable under a load that is uniformly distributed along its horizontal projection does into a parabola. Cables can be used in a variety of ways to span extremely large distances. Suspension cable structures and cable-stayed structures are two types of structures commonly used in building roofs, shown in Fig. 16-10.

Fig. 16-10　Cable Structures
(a) Suspension Cable Strucrures; (b) Cable-Stayed Sructures

(10) Arches: If the shape of a cable under a given load is inverted, the sag at any point is turned into a rise. The point is now above the line joining the end points, as shown in Fig. 16-11, by the same amount it is previously below it. A structure built according to the new shape is now in pure compression. A structure built according to the funicular shape is termed an arch, which belongs to line-forming structure. People in ancient times have constructed block arches made of separate pieces of brick and stone shown in Fig. 16-12. Rigid arches are frequently

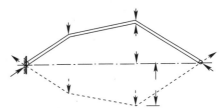

Fig. 16-11　Difference between Arch and Cable

Fig. 16-12　Block Arch

used in modern buildings, which are curved similarly to a block arch but made of a continuous rigid element such as steel or reinforced concrete member. The rigid arch is better to carry the variations of design loading than the block arch. Types of rigid arches are characterized by their support conditions shown in Fig. 16-13.

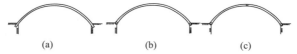

(a)   (b)   (c)

Fig. 16-13   Support Conditions of Rigid Arch
(a) Fixed; (b) Two-Hinged;(c) Three-Hinged

（11）Folded plate：A flat plate whose depth is small with respect to its in-plane dimension is typically used horizontally and carries loads by bending. Plates can be supported along their whole boundaries or only at selected points or with some mixture of continuous and point supports. Plate structures are normally made of reinforced concrete and steel. Long narrow rigid plates can be joined along their long edges and used to span horizontally. These structures, called folded plates, have higher load-carrying capacity than original flat plates. They are often used for building roofs shown in Fig. 16-14.

Fig. 16-14   Folded Plate Roof

（12）Shells：A shell is a thin rigid three-dimensional structure with a surface which may take any shape. Common forms are rotational surfaces generated by the rotation of a curve about an axis（e.g. spherical surface in Fig. 16-15a）, translational surfaces generated by sliding one plane curve over another plane curve（e.g. cylindrical surface in Fig. 16-15b）, ruled surface generated by sliding the two ends of a line segment on two individual plane curves（e.g. hyperbolic paraboloid surface in Fig. 16-15c）and a wide variety of complex surfaces formed by various combinations of the three types of surfaces. A shell is capable of supporting loads through compression, tension and shear stresses within its surface. The thinness of the surface does not allow the development of appreciable bending resistance. Thin shell structures are only suitable to carrying distributed loads and have found wide application as roof structures in buildings.

Three-dimensional forms may also be made of short rigid bars. Strictly speaking, these structures are not shell structures since they are not surface elements. However, their structural behavior is similar to continuous surface shells in which the stresses normally present in a continuous surface are concentrated into individual member. Structures of this type have been used extensively. Fig. 16-16 shows this type of structure, called a Schwedler dome, which comprises a lot of triangular meshes formed by bars.

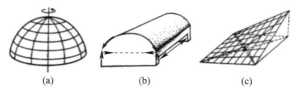

(a)   (b)   (c)

Fig. 16-15   Shell Structures
(a) Spherical Surface; (b) Cylindrical Surface; (c) Hyperbolic Paraboloid

Fig. 16-16　Schwedler Dome

（13）Membranes：A membrane is a thin flexible surface that carries loads primarily through the development of tension stresses. Soap bubble is a good example to illustrate what a membrane is and how it behaves. Membranes are sensitive to the aerodynamic effects of wind that can cause fluttering. Therefore, most of them used in buildings are stabilized in some way so that their basic shape is retained under various loading. A basic way of stabilizing a membrane is prestressing its surface. Fig. 16-17 and Fig. 16-18 show two kinds of membrane structures, called tent structures and pneumatic structures respectively. Prestressing can be achieved by means of the application of external forces that pull the surfaces taut for the tent structures, whereas by means of internal air pressurization for the pneumatic structures.

Fig. 16-17　Tent Structures

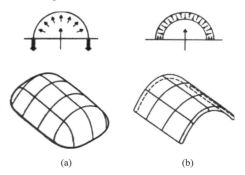

(a)　　　　　　　　　(b)

Fig. 16-18　Pneumatic Structures
(a) Air-Supported Form; (b) Air-Inflated Form

Pneumatic structures can be classified as air-supported structures and air-inflated structures. [6] The former is composed of a single membrane which is supported by the pressure of the internal air a little higher than that of atmosphere. The latter is supported by pressurized air contained within inflated building elements. The internal air of a building remains at atmospheric pressure. In both types of pneumatic structures, the air pressure induces tensile stresses in the membrane. External forces acting on the membrane cause a relaxation of some of these tensile stresses. The internal pressure must be high enough under the action of any possible applied loading to prevent compressive stresses from appearance.

Vocabulary and Expressions：

oft-times(ofttimes) ［'ɔfttaimz］ adv. ［古］时常，常常

waffle ［'wɔfl］ n. 华夫格子松饼，华夫饼筒

two-way slab 双向板

drop panel 无梁楼盖托板

enunciate ［i'nʌnsieit］ vt. 阐明，表明，规定

flat slab and columns 平板-柱

flat slab and shear wall 平板-剪力墙

coupled shear walls and beams 联肢剪力墙-梁

rigid frame 刚性框架

widely spaced perimeter tube 大柱距外框筒

rigid frame and haunch girders 刚性框架-加腋大梁

closely spaced perimeter tube 密布(小柱距)外框筒

perimeter tube and interior core walls 外筒内墙(筒中筒)

core supported structures 核心筒

shear wall-frame 框架剪力墙，框剪

shear wall-haunch girder frame 加腋梁框架-剪力墙

flat slab, shear walls and columns 平板-剪力墙-柱

frame tube structure 框筒结构

exterior diagonal tube 外部斜撑筒

modular tubes 成束(组合)筒

Notes：

① 这些因素包括建筑的几何形状，暴露于风的严重程度，该地区的地震活动，框架的延性，以及对结构构件尺寸的限制。

② 在混凝土建筑结构谱系中，最后一种是成束筒结构，它适用于超高层建筑和平面纵横比（即平面长宽比或平面宽高比）大的建筑（图 16-7）。

这里的"plan aspect ratio"翻译成"平面纵横比（即平面长宽比或平面宽高比）"，是指建筑平面图中房屋纵向长度与横向长度之比，与注⑤中的"height-to-width ratio"（建筑高宽比）不同，建筑高宽比是指建筑总高度与建筑平面图中的横向长度（即建筑宽度）之比。

③ 由于迎风和背风柱之间的巨大分隔，由此产生的系统就像一个巨大高效的垂直悬臂。

④ 例如，如果平面宽高比（或平面纵横比，平面长宽比）很大，比如说远远超过 1.5：1，很可能需要一个辅助的侧向支撑来满足漂移限制。

注意这里说的平面宽高比（或平面纵横比，平面长宽比）很大，是指平面高宽比（或平面横纵比，平面宽长比）1：x 中的 x 很大。所以翻译成 1.5：1 而不是 1：1.5。

⑤ 该建筑高 570ft（173.73m），高宽比异常高，为 8：1（图 16-6）。

⑥ 气动结构可分为空气支撑结构和充气结构。

Exercises：

1. Translate the Following Paragraph into Chinese.

Rigid frame system for resisting lateral and vertical loads have long been accepted as a standard means of designing buildings because they make use of the stiffness in the beams and columns that are required in any case to carry the gravity loads. In general, rigid frames are not as stiff as shear wall construction and are considered more ductile and less susceptible to catastrophic earthquake failures when compared to shear wall structure.

2. Translate the Following Paragraph into English.

一个好的建筑设计，需要一个好的结构形式去实现。而结构形式的最佳选择，要考虑到建筑上的使用功能、结构上的安全合理、艺术上的造型美观、造价上的经济以及施工上的可能条件，进行综合分析比较才能最后确定。

17 Connections and Construction of Steel Structure
（钢结构连接与施工）

教学目标：本单元的主要内容有钢结构连接与施工、螺栓连接、焊接以及高强度螺栓。教学重点是钢结构连接，难点是钢结构施工。通过本单元的学习，要求熟悉钢结构连接的类型和高强度螺栓，掌握螺栓连接和焊缝连接。

17.1　Steel Connections

17.1.1　Bolted Connections

For the last few decades, however, bolting and welding have been the methods used for making structural steel connections, and riveting is almost extinct. Bolting of steel structures is a very rapid field erection process that requires less skilled labor than does riveting or welding. This gives bolting a distinct economic advantage over the other connection methods in the United States where labor costs are so very high. Even though the purchase price of a high-strength bolt is several times that of a rivet, the overall cost of bolted construction is cheaper than that for riveted construction because of reduced labor and equipment costs and the smaller number of bolts required to resist the same loads. There are several types of bolts that can be used for connecting steel members.

The bolt connection makes the connecting pieces into a whole by the fastening force generated by bolts. According to the different bolt materials, it can be divided into ordinary bolt connection and high strength bolt connection.

According to the processing accuracy of bolts, common bolts are divided into three levels: A, B and C. Grade C, which has low processing accuracy and is matched with a hole diameter $1 \sim 1.5$mm larger than the bolt rod diameter, is generally used in steel structures. Therefore, it is suitable for the tensional connection along the rod axis (Fig. 17-1). Due to the large gap between the screw and the screw hole, when used for shear connection, there will be a large slip deformation after overcoming the friction between the connecting pieces, so it is only suitable for the connection

of secondary members (such as purlins, girts, bracing, platforms, small trusses and so forth) subjected to static loads or the installation connection with temporary fixed members (at this time the bolt is only used for positioning or clamping).

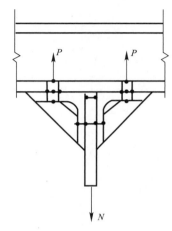

Fig. 17-1　Connection with tensile bolts

The strength of materials for high strength bolts is $3 \sim 4$ times of ordinary bolts. Enormous fastening pre-tensional force can be applied to high strength bolt screws, in order to make the connection plates subjected to magnificent compressive stress and fastened tightly. Shear force can be effectively transmitted by making use of the friction force between the plates. This connection type is called friction-type high strength bolt connection (Fig. 17-2). If the friction force between the connecting plates is not enough to bear loads and slip deformation occurs, shear force can be transmitted by the pressure force between the bolt rod and the wall of the screw hole. This kind of connection is called pressure-type high-strength bolt connection. High strength bolt connection can be widely used to install and connect important parts of steel structures.

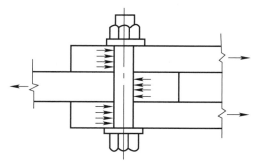

Fig. 17-2 Connection with shear bolts

17.1.2 Welded Connections

Welding is a process in which metallic parts are connected by heating their surfaces to a plastic or fluid state and allowing the parts to flow together and join (with or without the addition of other molten metal). It is impossible to determine when welding originated, but it was several thousand years ago. Metal-working, including welding, was quite an art in ancient Greece at least three thousand years ago, but welding had undoubtedly been performed for many centuries before those days. Ancient welding was probably a forging process in which the metals were heated to a certain temperature (not to the melting stage) and hammered together.

Although modern welding has been available for a good many years, it has only come into its own in the last few decades for the building and bridge phases of structural engineering. The adoption of structural welding was quite slow for several decades because many engineers thought that welding had two great disadvantages—(1) that welds had reduced fatigue strength as compared with riveted and bolted connections and (2) that it was impossible to ensure a high quality of welding without unreasonably extensive and costly inspection.

These negative feelings persists for many years, although tests seems to indicate that nei-

ther reason is valid. Regardless of the validity of these fears, they are widely held and undoubtedly slows down the use of welding—particularly for highway bridges and, to an even greater extent, for railroad bridges. Today most engineers agree that welded joints have considerable fatigue strength. They will also admit that the rules governing the qualification of welders, the better techniques applied, and the excellent workmanship requirements of the AWS (American Welding Society) specifications make the inspection of welding a much less difficult problem. Furthermore, the chemistry of steels manufactured today is especially formulated to improve their weldability. Consequently, today welding is permitted for almost all structural work other than for some bridges.

On the subject of fear of welding, it is interesting to consider welded ships. Ships are subjected to severe impactive loadings that are difficult to predict, yet naval architects use all-welded ships with great success. A similar discussion can be made for airplanes and aeronautical engineers. The slowest adoption of structural welding is for railroad bridges. These bridges are undoubtedly subjected to heavier live loads than highway bridges, larger vibrations and more stress reversals; but are their stress situations as serious and as difficult to predict as those for ships and planes? Several main advantages of welding are as follows:

(1) To most persons the first advantage is in the area of economy, because the use of welding permits large savings in pounds of steel used. Welded structures allow the elimination of a large percentage of the gusset and splice plates necessary for bolted structures as well as the elimination of bolt heads. In some bridge trusses it may be possible to save up to 13% or more of the steel weight by using welding.

(2) Welding has a much wider range of application than bolting. Consider a steel pipe column and the difficulties of connecting it to other steel members by bolting. A bolted connection may be virtually impossible, but a welded connection will present few difficulties. The

student can visualize many other similar situations in which welding has a decided advantage.

(3) Welded structures are more rigid because the members are often welded directly to each other. Frequently, the connections for bolted structures are made through intermediate connection angles or plates that deform due to load transfer, making the entire structure more flexible. On the other hand, greater rigidity can be a disadvantage where simple end connections with little moment resistance are desired. For such cases designers must be careful as to the type of joint they specify.

(4) The process of fusing pieces together gives the most truly continuous structures. It results in one-piece construction, and because welded joints are as strong as or stronger than the base metal, no restrictions have to be placed on the joints. This continuity advantage has permitted the erection of countless slender and graceful steel statically indeterminate frames throughout the world. *Some of the more outspoken proponents of welding have referred to bolted structures, with their heavy plates and abundance of bolts, as looking like tanks or armored cars when compared with the clean, smooth lines of welded structures.* ①

(5) It is easier to make changes in design and to correct errors during erection (and at less expense) if welding is used. A closely related advantage has certainly been illustrated in military engagements during the past few wars by the quick welding repairs made to military equipment under battle conditions.

(6) Another item that is often important is the relative silence of welding. Imagine the importance of this fact when working near hospitals or schools or when making additions to existing buildings. Anyone with close-to-normal hearing who has attempted to work in an office within several hundred meter of a bolted job will attest to this advantage.

(7) Fewer pieces are used and, as a result, time is saved in detailing, fabrication and field erection.

17.2   Steel Structure Construction

Construction of steel structures in a building starts with shop fabrication and handling. Then the components are transported to the site, erected into position and connected to form the structure. Since the in-place cost of steel in building structures ranges from two to four times the cost of steel as a raw material, ease of fabrication and erection is of prime significance to economy in construction. In order to reduce fabrication costs, one must use standard and available shapes as much as possible. While the variety and availability varies with the local situation, angles, plates and wide flanged shapes are the most easily obtainable. One should always try to pick sections that are readily available in the area where construction is taking place.

Having chosen possible sections to be used in a given construction, one would try to combine these shapes and sections to various subsystems that can be easily erected and connected. ②

The fabrication of these steel members and their connection is accomplished either by bolting or by welding. In earlier times, riveting was a common method. Riveting consists of a round steel bar forced in place to join several pieces of steel plates together. This method of connection and fabrication is now outdated by the use of high-tensile bolts. These high-tensile bolts depend on the clamping action produced when the nut or bolt is tightened to a predetermined tension. The second common method of connecting steel members is welding, which is a process applied by the use of heat to melt metals and to fuse components together. Briefly speaking, the heat is generated by an electric arc formed between a steel electrode and the steel part to be welded. The arc heat melts the base metal and the electrode simultaneously. The molten parts of the metal flow together, fuse, and are then allowed to cool off. Many techniques of shop and field welding have been developed.

The fabrication of standard shapes and plates into components ready for connection at the site is again an art in itself. Fabrication is generally carried out at the shop, using well-developed methods, tools and machinery, and employing mass-production techniques. To save labor and cost, one should minimize the types of shapes required for members and mechanize the fabrication process as much as possible. After fabrication of steel members is completed in a factory, structural components are transported to the site by railroad flat cars, trucks or barges. Then they are erected in a predetermined sequence, and they may require storage space. For multi-storey buildings, the components are usually erected in one-storey, two-storey or even three-storey tiers. For example, the World Trade Center is constructed of three-storey prefabricated columns and spandrel subassemblies. Thus after the foundations are completed, the columns are raised, set on the base plates and bolted in place. During erection, the columns are braced laterally until the structure is completed.

After the columns are set, beams and girders are hoisted in place and temporarily bolted to the columns. Industrial buildings of one or two storeys can be erected by small truck cranes. Buildings up to 200ft (61m) high are usually erected by larger truck cranes. Taller buildings require the use of special derricks, which are raised to the top of the completed portion of a structure as the building construction progresses upward. Self-lifting guy derricks are also used for multi-storey buildings. These derricks can jump themselves from one storey to another. The boom serves as a gin pole to hoist the mast to a higher level. The tower crane is more expensive than other types but has its own advantages. The control station can be located on the crane or at a distant position that enables the operator to see the load at all times. Also the equipment can be used to place concrete and other materials directly in the form for floors and roofs, eliminating chutes, hoppers and barrows.

Vocabulary and Expressions：

armor ['a:mə] n. 盔甲，装甲(车)
attest [ə'test] v. 证明(实)，表明
flexible ['fleksəbl] adj. 柔性的，可(易)弯曲的
forge [fɔ:dʒ] v. 锻造
inspection [in'spekʃən] n. 检查，检验
rigid ['ridʒid] adj. 刚性的，不易弯曲的
shank [ʃæŋk] n. 胫，末梢，后部
wrench [rentʃ] n. v. 扳手，拧
fabrication [ˌfæbri'keiʃn] n. 装配，制造，加工
bolt [bəult] n. 螺栓，螺杆，插销
weld [weld] n. v. 焊接，熔接，焊缝，焊点
rivet ['rivit] n. vt. 铆接，铆钉
clamp [klæmp] v. n. 夹紧，定位，夹子，夹具
fuse [fju:z] v. n. 熔合，保险丝
electrode [i'lektrəud] n. 电极，焊极，电焊条
mechanize ['mekənaiz] vt. 使机械化
barge [ba:dʒ] n. 驳船，平底船
spandrel ['spəndrəl] n. 拱肩，上下层窗空间
derrick ['derik] n. 人字起重机，摇臂吊杆
railroad flat car 铁路平板货车，铁路平板车
truck crane 汽车吊，汽车起重机
self-lifting guy derrick 自升牵索(栀杆)起重机
gin pole 起重桅杆，起重拔杆
gin pole derrick 栀杆起重机

Notes：
① 一些更直言不讳的焊接支持者提到了螺栓结构，与干净、平滑的焊接结构相比，螺栓结构有着厚重的钢板和大量的螺栓，看起来就像坦克或装甲车。
② 选好可用于某一结构中的截面形式之后，就要设法把这些型钢和各种形式的截面组合成各种容易安装和连接的分体系。

Exercises:

1. Translate the Following Sentences into Chinese.

(1) Even though the purchase price of a high-strength bolt is nearly three times that of a rivet, the overall cost of bolted construction is cheaper than that of riveted construction because of reduced labor and equipment costs and the smaller number of bolts required to resist the same loads.

(2) Welded joints are as strong as or stronger than the base metal.

(3) The important mechanical properties of most structural steels under static load are indicated in the idealized tensile stress-strain diagram shown in the figure.

(4) The values of E vary in the range $2 \times 10^5 \sim 2.1 \times 10^5 \, MPa$, and the approximate value of $2 \times 10^5 \, MPa$ is often assumed.

(5) Brittle fracture is initiated by the existence or formation of a small crack in a region of high local stress.

(6) High local stresses facilitate crack initiation, and so stress concentrations due to poor geometry and loading arrangements (including impact loading) are dangerous.

2. Translate the Following Sentences into English.

(1) 结构钢的延性取决于钢材的组成、热处理方法和钢材的厚度。

(2) 均匀厚度的翼缘便于铆接或焊接连接。

(3) 板梁就是由钢板、型钢通过铆接、焊接或螺栓连接组成的梁。

(4) 一般来说，焊接的板梁重量比铆接的轻。

18 Freeways
（高速公路）

教学目标：本单元的主要内容有高速公路设计速度、路肩宽度、标牌标记标线、互通立交以及匝道。教学重点是高速公路设计速度和标牌标记标线，难点是互通立交以及匝道。通过本单元的学习，要求熟悉标牌标记标线，了解互通立交以及匝道。

Freeways, especially those built to interstate standards are the safest of the various classes of highways. *While control of access, which limits vehicle conflicts, is a primary factor in relatively low accident, injury and fatality rates, other design features, such as wide medians and shoulders, roadsides clear of obstructions and the extensive use of protective barriers, are key factors as well.* [1] The higher design speeds used for freeways result in long sight distances due in long radius horizontal curves and long vertical curves, and other desirable design features that create a sale driving environment. Although most of the nation's freeways enjoy this relatively high level of design and safety, there are many opportunities for further enhancements. Safety improvements on freeways can also result in substantial savings in life and property because freeways carry 25% of the nation's total traffic.

Selecting the design speed for a freeway is an important safety element because most geometric criteria are related to or depend on it. In general, 90km/h should be the design speed for the mainline of a freeway, but it may need to be lower in areas of severe terrain or heavy development. *For reconstruction, rehabilitation and resurfacing (3R) projects, the design speed should not be less than the original design speed or the current legal speed limit of that highway section.* [2]

Design speeds for interchange ramps depend on the type of ramp selected, for example, loop, diamond or direct, and the low-volume, running speed of the intersecting highway. [3] Usually, the design speed is established by the most restrictive element of the ramps typically, the sharpest cane. Whatever design speed is selected, adequate transitions from the freeway proper and at the ramp terminal or merge point should be developed.

Safe and efficient traffic operations depend on adequate lane and shoulder widths outlined in *A Police on Geometric Design of Highways and Streets.* [4] The need to accommodate more traffic within existing or limited additional right-of-way on high-volume urban freeway has led some agencies to increase capacity by exchange full-lane or shoulder widths for additional travel lanes with reduced widths. Any proposed use of less-than-full-standard cross-sections must be analyzed carefully on a site-by-site basis. Experience indicates that 3.3m lane can operate safely if there are no other less-than-standard features; however, combined with shoulder-width reductions, substandard sight distance and other features, 3.3m lanes may not provide the same safe operation. Converting shoulders to travel lanes for additional capacity through a short bottleneck section has been shown to significantly reduce congestion-related accidents on some projects. Removing shoulders for several kilometers however has not had the same result.

Pavement markings, such as lane lines, edge lines, channelizing lines at interchanges and ramps and wont and symbol markings, provide important information to the driver. *Pavement markings define separate lanes of traffic traveling in the same direction, inform drivers of lane restrictions, and convey certain regulation and warnings that would not otherwise be clearly understood.* [5] Pavement markings are particularly important at night and during inclement weather and, therefore, must be retroreflective. While well-maintained painted pavement markings are acceptable, thicker, long-life markings, such as thermoplastic or preformed tapes, may perform better in wet weather and heavy traffic. Some studies have been made regarding the benefits of wider edge line markings (130 to 200mm versus the standard 100mm width). The wider markings may particularly benefit older driver.

Raised or recessed retroreflective pavement markers significantly enhance all types of standard markings, especially during heavy rain at night when standard markings are not readily visible. Hence, they are recommended for use on freeway whenever cost-justified. *Raised markers are placed on the pavement surface, and recessed markers are placed in shallow ground-out troughs.* [6]

It has been estimated that 8 billion hours of excess travel occur each year because of navigational errors that are partly due to inadequate signing, amounting to an average of 34 hours per year for every person in the United States. Appropriate highway information systems are needed to ensure that motorists have adequate time to acquire and process the information for control, guidance and navigation. On the other hand, the lack of signs or misleading signs can contribute to driver confusion, loss of attention and erratic maneuvers.

Signing for freeways should be planned concurrently with the geometric design. Trying to sign after interchange geometries and spacing have been finalized has often resulted in poor signing design and operational problem. How-

ever, poor freeway design cannot be overcome by signing. Freeway signing should be consistent, easy to read and unambiguous for the benefit and direction of drivers who are unfamiliar with the highway. In rural areas, the tendency to group all signing near the interchange should be avoided. Signs should be adequately spread on the approach to and departure from each interchange. They should be placed at natural target locations to command the driver's attention, especially where there is long distance between interchange and the alignment is relatively constant.

In areas with relatively close interchange spacing, the principle of sign spreading should be installed. Sign spreading helps disperse information so that motorists are not overloaded with a group of signs near the gore where path direction decisions are finalized. Fig. 18-1 shows an example of the sign-spreading concept. Using larger-than-standard letters can also help all drivers, particularly older drivers, in high-demand situations.

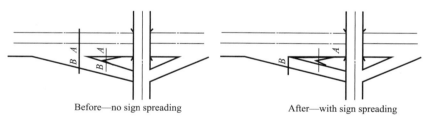

Before—no sign spreading   After—with sign spreading

Fig. 18-1 Sign Spreading

Where interchanges are so close together that major guide signs cannot be adequately spaced, they may be supplemented with interchange-sequence series signs that identify the next two or three interchanges. [7] If the interchanges spacing is less than 250m, then the series are used, it is preferable to use them over the entire length of the freeway or expressway within that urban area.

Advance signing for exits should be uniform, freeway-to-freeway splits, tangential freeway exit and interchanges with lane drops deserve special consideration. The *MUTCD* [8] pres-

ents additional guidelines to sign these situations. *Operational problems have occurred at freeway-to-freeway splits that feature a multilane exit and a second-route choice decision point on the exit ramp.* [9] Typically, there is no advance design for the second decision point due to spacing limitations. Consequently, the signs normally mounted over the second decision point are the first signs the drivers see indicating the appropriate lane for each direction on the destination freeway. Advance lane assignment is a signing concept that specifies the appropriate lanes for each direction on the intersecting freeway well before the decision point.

Vocabulary and Expressions：

wont［wəunt，wɔːnt］adj. 惯常的，习惯于；n. 习惯，惯常活动；v.（使）习惯于；动词原形 wont，过去式 wont，过去分词 wont 或 wonted，现在分词 wonting，第三人称单数 wonts

highway［ˈhaiwei］n. 公路

expressway［iksˈpreswei］n. 快速干道（部分立交），快速路

freeway［ˈfriːwei］n. 高速公路

trough［trɔf］n. 槽，水槽，饲料槽，海槽

erratic［iˈrætik］adj. 不稳定的，不规律的

maneuver［məˈnuːvə］n. 机动，调遣；v. 操纵，调遣

resurface［ˌriːˈsəːfis］vt. 罩面，重铺路面

rehabilitation［ˌriːhəˌbiliˈteiʃn］n. 重建，修复

ramp［ræmp］n. 匝道，斜坡道；v. 倾斜，做成斜坡；interchange ~ 立交匝道

loop［luːp］n. 环线，环道；v. 使成圈，循环

substandard［ˌsʌbˈstændəd］adj. 标准以下的，不合规格的；n. 低标准

inclement［inˈklemənt］adj. 残酷的，险恶的

retroreflective［ˌretrəuriˈflektiv］adj. 定向反光的，逆向反光的

thermoplastic［ˌθəːməuˈplæstik］adj. 热塑（性）的；n. 热熔物，热塑

recess［riˈses］n. 凹处，深处；vt. 使凹进

gore［gɔː(r)］n.（三角形的）端部，分道角；v. 将…剪成三角

accident (fatality) rate 事故（死亡）率

protective barriers 护栏，防护栏杆

sight distance 视距

low (high)-volume 低（高）交通量

travel lane 行车道

bottleneck section 瓶颈路段

pavement marking 路面标线

lanelines 车道线

edge lines 行车道边线

channelizing lines 渠化线

inclement weather 恶劣气候

raised or recessed retroreflective pavement markers 凸起或凹入式定向反光路面标线

guide signs 导向标志

advance signing 预告标志

Notes：

① 由于能减少车辆冲突，出入口控制是降低交通事故率和伤亡率的一个主要因素，但其他的设计特性，例如较宽的中央分隔带和较宽的路肩、路边无障碍物、大量使用防护栏等也是关键因素。

② 对于重建、修复和路面翻修工程而言，设计车速应不低于原有的设计速度，也不应低于该路段目前的法定速度限值。

③ 互通立交匝道的设计车速取决于所选匝道的类型，如环形、菱形还是定向式，也取决于相交道路的低流量的行车速度。

④ A Policy on Geometric Design of Highways and Streets, AASHTO, American Association of State Highway and Transportation Officials, 2011.《公路与城市道路几何设计》，美国国家道路和运输官员协会，2011 年。

⑤ 路面标线为同一方向交通流划分车道，告知驾驶员车道的限制，并传送用其他方法无法明确理解的管制信息和警告信息。

⑥ 凸起标记放置在路面上，凹入标记放置在浅挖地槽中。

⑦ 如果互通立交相距很近，主要的指路标志无法充分地间隔开，这时可采用立交排序标志以补充并确定前方的两个或三个立交。

⑧ MUTCD—Manual on Uniform Traffic Control Devices, FHWA, Federal Highway Administration, 2019.《统一交通控制设施手册》，美国联邦公路管理局，2009 年。

⑨ 在高速公路与高速公路立交的岔路口出现过操作问题，这些岔路口设有多车道出口，且在出口匝道处又存在二次选择路线问题。

Exercises：

1. Translate the Following Paragraph into Chinese.

Signing for freeways should be planned concurrently with the geometric design. Trying to sign after interchange geometries and spacing have been finalized has often resulted in poor signing design and operational problem. However, poor freeway design cannot be overcome by signing. Freeway signing should be consistent, easy to read and unambiguous for the benefit and direction of drivers who are unfamiliar with the highway. In rural areas, the tendency to group all signing near the interchange should be avoided. Signs should be adequately spread on the approach to and departure from each interchange. They should be placed at natural target locations to command the driver's attention, especially where there is long distance between interchange and the alignment is relatively constant. In areas with relatively close interchange spacing, the principle of sign spreading should be installed. Sign spreading helps disperse information so that motorists are not overloaded with a group of signs near the gore where path direction decisions are finalized.

2. Translate the Following Sentences into English.

（1）路基横断面形状取决于所采用的路面类型。

（2）路面标线为驾驶员提供重要的信息，在晚上及恶劣气候情况下，它们特别重要。

（3）无论采用什么设计速度，都应该设置从高速公路主线到匝道端点的过渡段。

（4）按设计标准修建的高速公路是各种等级公路中最安全的。

19　Cable-Stayed Bridge
（斜拉桥）

教学目标：本单元的主要内容有斜拉桥的构成、正交异性桥面板、斜索支撑、主墩以及桥塔。教学重点是斜拉桥的构成，难点是正交异性桥面板。通过本单元的学习，要求了解斜拉桥的构成、正交异性桥面板、斜索支撑、主墩以及桥塔等。

Cable-stayed bridge（Fig. 19-1）is constructed along a structural system which comprises an orthotropic deck and continuous girders which are supported by stays, i.e., inclined cables passing over or attached to towers located at the main piers. [①]

Fig. 19-1   Cable-Stayed Bridge

There are two major classes of cable-stayed bridges: In a harp design, the cables are made nearly parallel by attaching cables to various points on the tower(s) so that the height of attachment of each cable on the tower is similar to the distance from the tower along the roadway to its lower attachment. In a fan design, the cables all connect to or pass over the top of the tower(s).

The cable-stay design is the optimum bridge for a span length between that of cantilever bridges and suspension bridges. Within this range of span lengths a suspension bridge would require a great deal more cable, while a full cantilever bridge would require considerably more material and be substantially heavier. Of course, such assertions are not absolute for all cases.

19.1   History of Development

Cable-stayed bridges can be dated back to 1784, a design of a timber bridge by German carpenter C. T. Loescher. Many early suspension bridges are of hybrid suspension and cable-stayed construction, including the 1817 footbridge at Dryburgh Abbey, the later Albert Bridge (1872) and the Brooklyn Bridge (1883). Their designers find that the combination of technologies creates a stiffer bridge and John A. Roebling takes particular advantage of this to limit deformations due to railway loads in the Niagara Falls Suspension Bridge.

The earliest known example of a true cable-stayed bridge in the United States is E. E. Runyon's extant steel (or perhaps iron) bridge with wooden stringers and decking in Bluff Dale, Texas (1890). In the 20th century, early examples of cable-stayed bridges included A. Gisclard's unusual Cassagnes Bridge (1899), where the horizontal part of the cable forces was balanced by a separate horizontal tie cable, preventing significant compression in the deck, and G. Leinekugel le Cog's bridge at Lezardrieux in Brittany (1924). Eduardo Torroja designed a cable-stayed aqueduct at Tempul in 1926. Albert Caquot's 1952 concrete-decked cable-stayed bridge over the Donzére-Mondragon canal at Pierrelatte was one of the first of the modern type, but had little influence on later development. The steel-decked bridge designed at Strömsund by Franz Dischinger (1955) is therefore more often cited as the first modern cable-stayed bridge.

Other key pioneers included Fabrizio de Miranda, Riccardo Morandi and Fritz Leonhardt. *Early bridges from this period used very few stay cables, as in the Theodor Heuss Bridge (1958). However, this involves substantial erection costs, and more modern structures tend to use many more cables to ensure greater economy.* [2]

19. 2　Comparison with Suspension Bridge

A multiple-tower cable-stayed bridge may appear to a suspension bridge (Fig. 19-2), but in fact is very different in principle and in the method of construction. In the suspension bridge, a large cable is made up by "spinning" small diameter wires between two towers, and at each end to anchorages into the ground or to a massive structure. These cables form the primary load-bearing structure for the bridge deck. Before the deck is installed, the cables are under tension from only their own weight. Smaller cables or rods are then suspended from the main cable, and used to support the load of the bridge deck, which is lifted in sections and attached to the suspender cables. As this is done the tension in the cables increases, as it does with the live load of vehicles or persons crossing the bridge. The tension on the cables must be transferred to the earth by the anchorages, which are sometimes difficult to construct due to poor soil conditions.

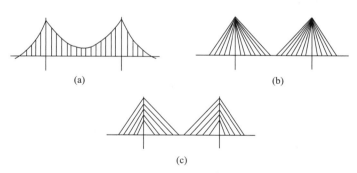

Fig. 19-2　Suspension Bridge and Cable-Stayed Bridge
(a) Suspension Bridge; (b) Cable-Stayed Brige, Fan Design; (c) Cable-Stayed Brige, Harp Design

In the cable-stayed bridge, the towers form the primary load-bearing structure. A cantilever approach is often used for support of the bridge deck near the towers, but areas further from them are supported by cables running directly to the towers. This has the disadvantage, compared to the suspension bridge that the cables pull to the sides as opposed to directly up, requiring the bridge deck to be stronger to resist the resulting horizontal compression loads; but has the advantage of not requiring firm anchorages to resist a horizontal pull of the cables, as in the suspension bridge. *All static horizontal forces are balanced so that the supporting tower does not tend to tilt or slide, needing only to resist such forces from the live loads.* ③

Key advantages of the cable-stayed form are as follows:

(1) Much greater stiffness than the suspension bridge, so that deformations of the deck under live loads are reduced.

(2) Can be constructed by cantilevering out from the tower—the cables act both as temporary and permanent supports to the bridge deck.

(3) For a symmetrical bridge (i.e., spans on either side of the tower are the same), the horizontal forces balance and large ground anchorages are not required.

A further advantage of the cable-stayed bridge is that any number of towers may be used. This bridge form can be as easily built with a single tower, as with a pair of towers. However, a suspension bridge is usually built only with a pair of towers.

Vocabulary and Expressions:

assertion [ə'sə:ʃən] n. 断言，主张
footbridge ['futbridʒ] n. 人行桥，人行天桥
stringer ['striŋə(r)] n. （铁路的）纵梁，纵桁，纵向轨枕
aqueduct ['ækwidʌkt] n. 水管，沟渠，导管
spinning ['spiniŋ] n. 纺纱，旋转，自转
harp design 竖琴式设计
orthotropic deck 正交异性桥面板
extant [ek'stænt] adj. 现存的，显著的
a massive structure 又大又重的结构
suspender cable 吊索

Notes:

① 斜拉桥是依照这样的结构体系建造而成：该体系由正交异性桥面板（用纵横向互相垂直的加劲纵肋和横肋连同桥面盖板所组成的共同承受车轮荷载的结构）和斜索支撑的连续梁组成，即倾斜的缆索跨越或连接在主墩的桥塔上。

② 早期的斜拉桥只用很少的斜索，比如1958 年建造的 Theodor Heuss 桥；然而，这种形式涉及高昂的施工造价，而更现代一点的结构则趋向于使用较多的斜索以确保更好的经济性。

③ 所有的静水平力都可以平衡掉，所以主塔不会出现倾斜或滑移，主塔只需要抵抗活载产生的力。

Exercises:

1. Translate the Following Paragraph into Chinese.

Hangzhou Bay Bridge is a bridge with cable-stayed bridge portion across Hangzhou Bay off the eastern coast of China, which linked up on June 14, 2007. It connects the municipalities of Shanghai and Ningbo in Zhejiang province. The bridge is the longest trans-oceanic bridge in the world, although it does not have the longest cable-stayed main span. The opening ceremony was held on June 26, 2007 with great domestic media publicity, though after the opening ceremony, the bridge would only be used for test and evaluation purposes. It was opened to the public on May 1, 2008.

2. Translate the Following Sentences into English.

(1) 在过去的几十年间，斜拉桥已经得到了广泛的应用。

(2) 在斜拉桥设计中，需要考虑桥梁与周围环境相协调。

(3) 由于它们所具有的结构特性，斜拉桥的跨径介于梁式桥与悬索桥之间。

20　Tunnel Engineering

（隧道工程）

　　教学目标：本单元的主要内容有隧道类型、导洞隧道、明挖隧道、软基隧道、硬岩隧道、双圆盾构隧道、隧道掘进机、铁路隧道、公路隧道、地铁、水力发电站以及新奥隧道施工法。教学重点是隧道类型、硬岩隧道和盾构隧道，难点是新奥隧道施工法。通过本单元的学习，要求熟悉隧道类型和隧道掘进机，了解新奥隧道施工法。

　　A tunnel is a long, narrow, essentially linear excavated underground opening, the length of which greatly exceeds its width or height. For centuries, mankind has excavated tunnels in the earth for a myriad of uses. With improvements in design and construction, the diversity of uses has also widened, and tunnels are not simply the mines and shelters they used to be. Nowadays a great deal of countries are using tunnels to provide spaces for human activities such as living, storage, communication, power transmission and transportation.

　　Tunnels are one of the greatest projects in engineering and construction. It is also one of the most expensive projects. For this reason, extensive planning and surveying goes into the pre-excavation stage of the project.

20.1　Types and Structures of Tunnels

20.1.1　Tunnel Types

　　The response of the ground to the excavation of an opening can vary widely. Based on the type of ground in which tunnelling takes place, four principal types of tunnelling may be defined:

　　(1) *Cut- and-cover tunnel*[①] is shown in Fig. 20-1. In such tunnels, the ground acts only passively as a dead load applied to a tunnel which is to be erected like any aboveground engineering structure.

Fig. 20-1　Cut-and-Cover Tunnel

　　(2) *Soft ground tunnel*[②] is shown in Fig. 20-2. In such tunnels, immediate supports must be provided by a stiff lining. The ground usually participates actively by providing resistance to the outward deformation of the lining.

　　(3) *Medium-hard rock tunnel*[③] is shown in Fig. 20-3. In such tunnels, the tunnels are con-structed in medium-hard rocks or in more cohesive soils, where the ground may be strong enough to allow a certain open section at the tunnel face. A certain amount of stress release may permanently be valid before the supporting elements and the lining begin acting effectively. In this situation only a fraction of the primary

ground pressure is acting on the lining.

Fig. 20-2　Soft Ground Immersed Tunnel

Fig. 20-3　Multi-Face Tunnel in Medium-Hard Rock

(4) *Hard rock tunnel*[4] is shown in Fig. 20-4. In such tunnels, the hard rock ground may preserve the stability of the excavation so that only a thin lining will be necessary.

Fig. 20-4　Hard Rock Tunnel

20. 1. 2　Tunnel Structures

In the design of tunnels, designers attempt to utilize a structure that will prove to be highly effective. Basic mechanics in conjunction with the character of rocks or soils, can dictate a most effective structure. For instance, tunnels are rarely excavated with a flat roof. The reason is that, as span increases, the rock or soil in the center part of the span has less force holding it up, and the flat-roofed tunnel is more susceptible to collapse. Therefore, for the most part, tunnels are excavated with circular arched roofs, which are the most stable structures with regard to externally applied loads. For this reason tunnels are commonly constructed with circle shaped, horseshoe shaped sections or with *Gothic arch roofs*[5], which provide maximum stability.

In addition to the shape of tunnel sections, the structure and size are also very important. Doubling the diameter of a tunnel requires removing four times as much soil or rock, the surface area of the tunnel is doubled and the forces of the soil or rock are now over twice. So the use of supporting structures is much more important in large tunnels than in smaller ones. Often, in cases where tunnels are to be driven into rock of questionable competence, very small tunnels are driven first and then carefully enlarged and supported as shown in Fig. 20-5.

Fig. 20-5　Pilot Tunnel in Medium-Hard Rock

With respect to shield-driven tunnels often used in soft soils, the structures of the tunnels are as follows:

(1) Circle structures: This type of shield-driven tunnel is widely used in constructing metros in soft ground and has a number of advantages compared with other types of structures. First and foremost, the circle structure is more economical and very convenient in construction. The driving shield machines are easy to be made and operated and more reasonable in distribution of internal forces compared with other structure shapes.

(2) Rectangular and horseshoe structures:

This type of structure is not widely adopted since it is more complicated and expensive. Furthermore, the distribution of internal forces is not simple either.

(3) *DOT*[6] structures: The DOT structure shield came to Shanghai and was used in the construction of metros line M8 in 2002. Fig. 20-6 illustrates the structure of it.

Fig. 20-6   The DOT Structure Shield

20. 2   Construction of Tunnels

20. 2. 1   Railway Tunnels

Among the 5300 railway tunnels in China, most of them have been constructed using the *NATM* (*New Austrian Tunnelling Method*) *method* (*Fig.* 20-7).[7] NATM is firstly adopted in the Xiaken Railway Tunnel of Wanggang Railway. The tunnel with a length of 65m and a depth of 20m is constructed in heavily weathered strata, where rocks are not only weak and fractured but also rich in groundwater. On the basis of the principles of NATM, the tunnel is excavated in a large cross section by use of presplit blasting or smooth blasting for minimizing disturbance to the surrounding rock. After excavation, bolts and shotcrete are immediately mounted to control the deformation of the surrounding rock and a secondary concrete lining is mounted when the deformation of the surrounding rock is basically stable. During the construction, the surrounding rock is monitored and the construction is adjusted according to the monitoring results. The successful experience of NATM in the Xiaken Railway Tunnel has been introduced rapidly to other railway tunnels. At present, NATM has become the most popular principle on the design and construction of tunnels in China. Chinese engineers have built considerable tunnels in difficult geology by means of NATM, which has enriched their experience of using NATM. For example, the construction of the tunnels on Nanning-Kunming railway which was opened into operation in 1997 is a typical example in unfavorable geology. There are 258 tunnels with a total length of 194. 6km that have used NATM. During the construction of the tunnels, difficulties, such as weak surrounding rock, high geostress, shallow depth, uneven pressure, karst, water inflow, faults, concerted gas and high seismic intensity, have been overcome one after another. The completion of Nanning-Kunming Railway demonstrates that China has capability of building tunnels in almost all unfavorable conditions of geology.

Although a large number of railway tunnels have been built in China, by use of drill and blast except the Qinling Tunnel which is constructed by TBM method, it is clear that other construction methods such as shield method and the method of immersed tunnels shall also be adopted for building some railway tunnels in special geology.

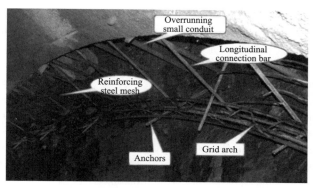

Fig. 20-7   NATM Method

20. 2. 2   Highway Tunnels

On the basis of the experience drawn from the construction of railway tunnels, the highway department has successfully designed highway tunnels with large cross sections and underwater and has accumulated much experience on the construction of these tunnels.

（1）Highway tunnels with large cross sections: Building three lane highway tunnels is unavoidable with the construction of six-lane expressway. In China, several tunnels with three lanes are finished, which are Shanxinpo Tunnel in Yunnan, Kaoyishan Tunnel in Guangzhou, Tieshanping Tunnel and Zhenwushan Tunnel in Sichuan, Tanyugou Tunnel in Beijing, etc. The highway department has conducted several research projects on the calculation of structures, excavation and support measures of three-lane tunnels and has obtained a number of useful research findings which have been used in the construction of the above tunnels.

（2）Highway tunnels underwater: On January 8, 1994, the first immersed tunnel in China, i.e., the Zhujiang Immersed Tunnel in Guangzhou, was put into operation, which was designed and built by Chinese companies. The tunnel with a width of 33m and a height of 7. 96m is 1239m in the total length including 517. 5m approaches on both banks and 264m is bored by the use of mining method and 457m by use of the immersed method. It consists of two chambers for highway, with two lanes for each,

one chamber for metro and one small chamber for cables. With regard to the immersed method in detail, the first section of the tunnel will probably be placed on the one side of the site, but before this can be undertaken, a great deal of preparatory work has to be done. During the prefabrication of the tunnel elements the site entrance portal will be constructed. A trench long enough to receive tunnel elements will be dredged from the side of the site. When this is completed the first element can be maneuvered over the trench and is ready to be gently lowered into position against the site portal. The lowering operation requires a high degree of precision to avoid making damage to the elements and to ensure that a reliable water tight seal is made with the portal or preceding tunnel elements. Finally, the trench is infilled and topped with a layer of rock armor to provide protection for the tunnel.

Construction operations are generally carried out underwater from surface vessels which require an uninterrupted working area for maneuvering. The operations are time consuming, because at each stage accurate positional checks have to be made. If one-side sections are placed on the site first, the work will then move over to the other side where the remaining sections will be placed and linked together each other. On August 8, 1995, the second immersed tunnel in China, i.e., the Yongjiang Immersed Highway Tunnel in Zhejiang Province, was put into use. The tunnel has a total length of 1019. 53m and a length of 420m in the immersed section. The construction of those two immersed highway tunnels demonstrates that the techniques of immersed tunnel construction have been accepted in China. A final transportation concept for crossing straits, fiords and lakes is a submerged floating tunnel as shown in Fig. 20-8. The submerged floating tunnel is a natural way to bridge water and utilizes lakes and waterways to carry traffic under water and on to the other side, where it can be conveniently linked to the rural network or to the underground infrastructure of modern cities.

Fig. 20-8　The Submerged Floating Tunnel

20.2.3　Metros

The first section of the metro in China was built in Beijing, which was opened to traffic in the late 1960s. After that, the first section of metros in Tianjin, Shanghai and Guangzhou was put into use in 1980, 1994 and 1998 respectively. During the construction of metros in Beijing and Tianjin between the 1960s and 1980s, the cut and cover method was widely adopted owing to scattered buildings in these cities at that time. However, the cut and cover method makes citizens' lives inconvenient, though it has advantages of low cost and high advance rate.

Between the 1980s and 1990s, due to the development of the cities, the shield method and the mining method was adopted for the construction of metros so as to reduce the disturbance to surface traffic and citizens' lives and to avoid the relocation of surface buildings, underground cables and pipes. During the construction of the

first line of the Shanghai metro (Fig. 20-9), seven shields were imported from aboard for the driving of running tunnels under busy streets or high buildings. The advance rate of the shields is 4~6m per day on an average, 130m per month on an average and 320m per month in maximum. In the early 1990s, the shield method was also used in the construction of the running tunnels on the first line of Guangzhou metro.

Fig. 20-9　Shanghai Metro Constructed by Shield Method

Besides the shield method, NATM was widely adopted for the construction of both running tunnels and stations in the Fuxingmen—Bawangfeng line of Beijing Metro and the first line of Guangzhou Metro between the 1980s and 1990s. NATM is still the most important method used in the construction of metros in China. In the construction of stations, the application of cut and cover method has increased in recent years in China. The cut and cover method was used in the construction of three stations in the first line of Shanghai Metro in the early 1990s and was later adopted for the construction on Fuxingmen—Bawangfeng line of Beijing Metro and the first line of Guangzhou Metro.

20.2.4　Hydraulic and Underground Powerhouses

In the construction of hydraulic tunnels and underground powerhouses, there are three example projects in China:

(1) Tianshengqiao hydro-power station built in the 1980s, includes three division tunnels, each with a diameter of 10.4m and a length of

9.78km, which were constructed in the strata of developed karst. In order to speed up the advance rate of excavation, two second-hand tunnel boring machines (TBMS) with a diameter of 10.8m were imported from aboard (Fig. 20-10). Due to unfavorable geology, such as karst debris flow and underground rivers, the construction went slowly and was completed behind schedule.

Fig. 20-10 Tunnel Boring Machine (TBM)

（2）Gansu Yindaruqin division project was constructed in the early 1990s, with a length of 9.649km and a diameter of 5.53m. By use of double shield TBM, the construction of the tunnel was completed in 13.5 months. With a best daily progressing rate of 65.5m and a best monthly rate of 1300m, the TBM method has aroused general interest in China. After that, the Ministry of Railways started to consider the possibility of using TBM to excavate a railway tunnel.

（3）Shanxi Wangjiazhai Yellow River division project is under construction. The project consists of main section, south section and north section, including 192km long tunnels. The longest tunnel, the No.7 tunnel, is up to 43km long and is constructed by using TBM.

In the construction of hydraulic and underground powerhouses, the underground powerhouse of Liujiaxia hydro-power station built in the 1960s and was 24.5m in width and 62.5m in height. In the 1980s and 1990s, several underground powerhouses were built in Baiyun, Lubuge, Dongfeng, Guangzhou pumped storage and Shisanling pumped storage power stations. Fig. 20-11（a）and（b）illustrate two hydro-power stations.

(a)　　　　　　　(b)

Fig. 20-11 Hydro-Power Station
(a) Liujiaxia Hydro-Power Station; (b) Three Gorges Power Station

20.3　Vision of the Future

Due to the high level of economic development and huge investments in infrastructures in China, the construction of extra-long tunnels and large underground works will be needed in the future. Key future projects include the Qiongzhou Strait Tunnel in Hainan, the crossing project of bridges and tunnels in Bohai Gulf with some highway tunnels longer than 10km, metros in more than 20 cities, etc. The construction of these projects not only brings new opportunities for the tunnelling industry, but also presents a challenge to engineers in China. It is believed that the tunnelling technique in China will have great advances in the near future.

Vocabulary and Expressions：
pilot tunnel 导洞隧道
shield-driven tunnel 盾构隧道
TBM（Tunnel Boring Machine）隧道掘进机

Notes：
① 大开槽施工法隧道（随挖随填隧道），即明挖法隧道。
② 软基隧道
③ 中等硬度岩石隧道。
④ 硬岩隧道。
⑤ 哥特式尖拱顶。
⑥ DOT（Double-O-Tube）双圆潜盾机，Double-O-Tube shield tunne 双圆盾构隧道。
⑦ NATM（New Austrian Tunnelling Method）

新奥隧道施工法。

Exercises：

1. Answer the Following Questions in English.

（1）What is a tunnel?

（2）What are the main types of tunnels?

（3）Explain, with examples, the difference between a rock tunnel and a soil tunnel.

（4）How important do you consider environmental design as a part of the tunnel design program?

2. Translate the Following Paragraph into Chinese.

New Austrian Tunnelling Method "NATM" is the application of the theory of rock mechanics, that is, after tunnel excavation, bolt and shotcrete is adopted as the main timely support means, in order to maintain and use the bearing capacity of surrounding rock itself, control the deformation and relaxation of surrounding rock, and make surrounding rock becoming a part of the support system. The methods and principles to guide the tunnel and underground engineering design and construction are obtained by the measurement and control of surrounding rock and support. The design of tunnel is based on standard support, information feedback and data analysis, so the quality control of initial support system should be paid much attention to and strengthened in construction.

3. Translate the Following Paragraph into English.

隧道是修建在地下、水下或山体中，铺设铁路或修筑公路供机动车辆通行的建筑物。根据其所在位置可分为山岭隧道、水下隧道和城市隧道三大类。为缩短距离和避免大坡道而从山岭或丘陵下穿越的称为山岭隧道；为穿越河流或海峡而从河下或海底通过的称为水下隧道；为适应铁路通过大城市的需要而在城市地下穿越的称为城市隧道。这三类隧道中修建最多的是山岭隧道。

21 Characteristics of China's Imperial City Planning

（中国皇城规划特征）

教学目标：本单元的主要内容有中国皇城规划特征、瓮城和马面、巷和胡同、风水和堪舆、四周围合以及里坊制。教学重点是中国皇城规划特征，难点是风水和堪舆。通过本单元的学习，要求掌握中国皇城规划的十一个特征，了解中国皇城规划中的瓮城和马面、巷和胡同、风水和堪舆、四周围合以及里坊制等。

Basic Feature: a Four-Sided Enclosure. The basic feature of China's imperial city is a four-sided enclosure (Fig. 21-1). Every Chinese imperial city is encased by four outer walls which meet at right angles to form a rectangle. Within the walls are at least one and sometimes two or more sets of walls that define smaller rectangular enclosures. Inside the smaller enclosures palatial sectors are elevated. Until the 14th century the material of outer enclosures was most often pounded earth. In later times outer walls were sometimes faced with brick. Smaller enclosures were generally constructed with plaster walls or pillared arcades. The most extensive enclosures of all, the Great Wall of China, was built of pounded earth beginning in the first millennium B.C.

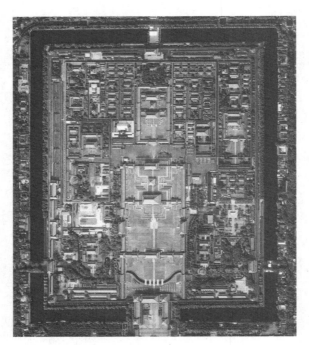

Fig. 21-1　Map of the Palace Museum

Outer and Inner City Walls Pierced by Gates: The second characteristic of the Chinese imperial city is that outer and inner city walls are pierced by gates. Often, but not always, three gates are found at each outer wall face for a total of twelve. Ideally, gates of opposite city walls are equidistant from the adjoining wall corners. Inner cities of the imperial capital generally have no more than one gate at each side. One central gate always is placed at a south city wall. Most city wall gates are built for entrance or exit by a land route, usually a major urban thoroughfare, but some imperial cities have sluice gates as well.

Outer Walls Possessing No Defensive System: The third feature of China's imperial city outer walls is the defensive projection, which take the form of a lookout tower or a protective battlement. Lookout towers are built at the four corners of a city and atop city gates, where troops can be quartered. The two types of battlements most common in Chinese imperial cities are wengcheng and mamian（瓮城和马面）. Wengcheng are additional walls built in front of gates. They projected in front of and up to outer wall gates, with their own openings for access to the city（Fig. 21-2）. Mamian, literally "horse faces", alternately known as yangma（羊马）, are simply additional fortified perimeter space that curved around the outer city wall at intervals, providing no access to the city interior. The use of mamian in medieval China is confirmed by an illustration in the 11th-century military Wujing zongyao（《武经总要》Collection of Important Military Techniques）（Fig. 21-3）. Both defensive wengcheng and mamian are also used at nonimperial Chinese cities.

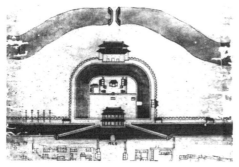

Fig. 21-2 Wengcheng of Anding Gate
（安定门）in Beijing

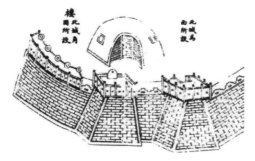

Fig. 21-3 Mamian (Yangma)

Orthometric Streets Forming Clearly Articulated and Directed Space：The plan of Imperial Beijing（Fig. 21-4）shows that major north-south and east-west streets crossed the city at right angles. *The city roads and avenues often run from a northern to a corresponding southern*

gate, or from an eastern to a western one, giving way to a design feature that can be called clearly articulated and directed space. [1] The principal north-south thoroughfare runs along a line that passes through the central northern and southern gates of each city wall. Parallel to the gate-initiated streets in both directions are smaller avenues, and parallel to them are xiang and hutong（巷和胡同）, east-west and north-south oriented lanes and alleys. The unambiguous articulation of north-south and east-west space is such that even the city's smallest regions are encased by walls or streets that run perpendicular to one another. Seen from above, the city of Beijing appears as a checkerboard in maps drawn during the reign of Qianlong emperor（1735~1795）, and the image is still apparent in the 20th-century photograph of Xi'an, formerly an imperial city. Orientation of the streets, like the walls, is according to the four cardinal directions—symbolically the clearly demarcated boundaries of the Chinese empire.

Orientation and Alignment：The four-sided Chinese city is a physical manifestation of the traditional belief in a square-shaped universe, bounded by walls, with the Son of Heaven at its center. Tradition associates each of the four world quarters and the center with a symbolic animal, color, metal, season（excluding the center in this case）and a host of other phenomena. South, for instance, is the direction of summer, fire, the bird（often a phoenix）and the color vermilion. South is the cardinal direction the emperor faced when seated in his hall of audience, and thus most of the imperial buildings of an imperial city have a southern exposure. Continuing around the square, east is the quadrant of spring, wood and the azure dragon（青龙）; north is winter, water and the black tortoise; west is autumn, metal and the white tiger. Autumn and white are associated with death in Chinese culture, fall being the season of decay that, after the freezing of winter, will give way to new life and renewal in the spring. Thus, tombs are often constructed north or west of the capital, the quadrants of death and

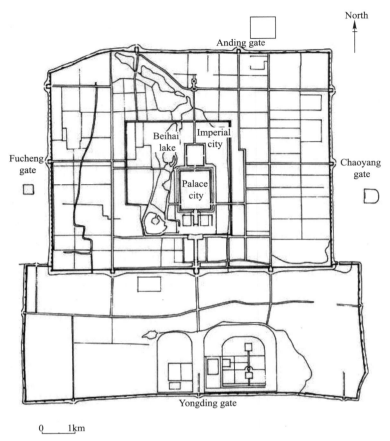

Fig. 21-4   Plan of Imperial Beijing

decay. The names of imperial city structures also reflect attitudes toward cosmological alignment. At many Chinese imperial cities the palace of the crown prince or another hall is named Taiji（太极）, a reference to the polar star, and the Forbidden City of Beijing is known as Zijin Cheng, the Polar Forbidden City.

More Enclosures of Almost Every City Sector Being Easier for Government to Manage：*Four-sided enclosure of the city and cardinal orientation of its major routes lent themselves to the further enclosure of virtually every city sector according to the four cardinal directions.*[②] Bounded regions of the city made census taking and population control possible even in the late first millennium B. C. By the time of the 7th through 9th centuries a sophisticated system of one hundred and eight walled wards was in place in the capital Chang' an（Fig. 21-5）. Just a cen-

tury later the ward system（里坊制）would be a weak reflection of its former self, but the practice of dividing the city into governmentally controlled spaces that were inhabited predominantly by people of one occupational, religious or ethnic group persisted in Beijing into the 20th century.

Accessibility of Water：Easy access to a good water supply is also essential to the plan of every Chinese imperial city. In addition to choosing a site near an abundant water source, the Chinese capital is usually surrounded by a moat, and human-directed waterways are channeled into the city. The Three Lakes of Beijing（Rear Lake, Shichahai Lake and Beihai Lake）are examples of artificial water sources dug in the 12th century that is preserved by each dynasty.

Vast Size：Vast size may also be considered characteristic of the Chinese imperial city. Until Beijing in the 15th century, most Chinese

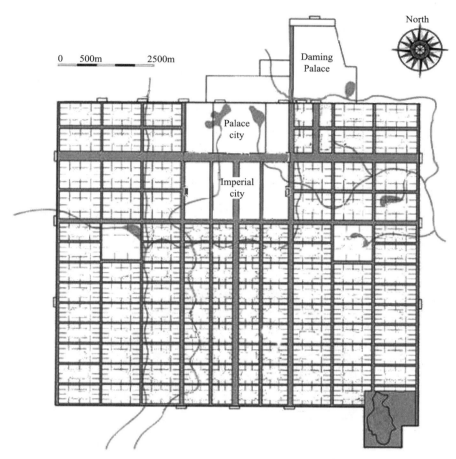

Fig. 21-5   Plan of Sui-Tang Chang'an Showing Its One Hundred and Eight Wards

capitals were the largest cities in the world during their times of flourishing. Beijing shown in Fig. 21-4 encompassed an area of 62km². In the 7th century the outer wall of the capital Chang'an spanned a distance of 36.7km. In the Warring States period （475 ~ 221 B. C.) the capital of just one state, Yan, was about 32km².

Huge Population： Related to the size of the Chinese imperial city is another feature： its population. Until the 14th or 15th century most primary Chinese capitals had the largest urban populations in the world. *Means of assessing the population of traditional Chinese cities vary from strict reading of local records to educated interpolations of them, and opinions differ about the validity of counting resident military in urban population figures.* [3] It is generally agreed, however, that both Chang'an in the 8th century and Beijing at the end of the 16th century had populations of one million. The 10th through 13th centuries Song dynasty capitals at modern-day Kaifeng and Hangzhou had populations of well over one million; even the early Chinese capitals of Chang'an in the last centuries B. C. and Luoyang in the first and second centuries A. D. had at least three hundred thousand people.

One reason for the huge concentration of people in Chinese capitals is the common imperial practice of relocating masses of the population. [4] If historical texts can be trusted, then at times hundreds of thousands of people are transferred to new capital cities. The lower echelons of society functions as builders of new imperial projects, the upper classes are state servants, and those in the middle serves as merchants, artisans and commercial agents for the newly transplanted urban population.

Siting: Another characteristic of Chinese imperial city may be called siting. Siting describes the belief that natural phenomena—mountains, wind, water—must be harmoniously interrelated at a site in order to ensure auspicious human existence. The practice of divination for the purpose of determining a positive balance of natural forces before selecting a site is often called Fengshui or Kanyu (风水和堪舆), both sometimes translated as Chinese geomancy. References to the process by which a site is chosen are found in early Chinese written and illustrative records (Fig. 21-6). Aspects of siting that have traditionally been concerns for imperial city builders are the location of protective mountains (or in their absence an artificial hill) to the capital's north and water to the south.

Building Order: The final characteristic of Chinese imperial city planning has not always been possible to implement, but it should be mentioned. Ideally Chinese imperial city is planned in entirety from its inception and is constructed beginning with the outer wall. *At times building inside the outer wall occurred before the enclosure is complete, but the size and shape of the wall, and thus the enormous size of imperial cities in China, are rarely accidental.*[5]

Fig. 21-6　The Ancient Compass Choosing "a Site" for a Dwelling

Vocabulary and Expressions:

urban-rural planning 城乡规划

Chinese Imperial City Planning 中国皇城规划

palatial [pə'leiʃəl] adj. 宫殿似的，宏伟的

pounded earth 夯土

plaster ['plɑːstə(r)] n. 石膏，灰泥

pillared arcade [ɑː'keid] 带柱拱廊

equidistant [ˌiːkwi'distənt, ˌekwi'distənt] adj. 等距的，距离相等的

thoroughfare ['θʌrəfɛə(r)] n. 大道，要道

sluice [sluːs] n. 水闸，水门

demarcate ['diːmɑːkeit] vt. 划分界线，区别

vermilion [və'miliən] n. [无化]朱砂，朱红色；adj. 朱红色的；vt. 涂以朱红色

azure ['æʒə(r), 'æzjuə(r)] dragon 青龙

diviner [di'vainə] n. 占卜者，占卜师，预言师

echelon ['eʃəlɒn] n. 阶层，梯队

geomancy ['dʒiːəumænsi] n. 风水，堪舆

Notes:

① 皇城的街道一般都起止于相对的南北门或东西门，而形成空间结构清晰、导向明确的设计特点。

② 皇城四周围合以及基于四个基准方位布

置主要道路的特点，使几乎每个城市单元都能进一步基于四个基准朝向进行围合。

③ 中国传统城市的人口统计方法各不相同，有严谨的地方记录，也有对原始数据的精心增补。而对城市人口中驻扎军队数量的计算方法，其可靠性人们也有不同看法。

④ 人口大量聚集的原因之一是政府有关人口迁移和管理方面的政策。

⑤ 有时候，内城建设在外围城墙完工之前就已开始，但是城墙的尺寸和造型，以及相应形成的皇城的巨大尺度，大多也是出于精心的设计。

Exercises：

1. Fill in Each of the Blanks with the Correct Answer.

（1）Outer and inner city walls were pierced by gates, and ideally gates of opposite city walls were_____ from the adjoining wall corners.

A. equidistant   B. equivalent
C. commensurable   D. comparable

（2）The defensive projection of Chinese imperial city outer walls often took the form of a lookout tower or a _____ , such as wengcheng and mamian.

A. protective battlement   B. safeguard wall
C. protective artillery   D. strategic castle

（3）The unambiguous articulation of north-south and east-west space is such that even the city's smallest regions are encased by walls or streets that run _____ to one another.

A. parallel   B. correspondent
C. across   D. perpendicular

（4）The four-sided Chinese city is a physical manifestation of the traditional belief in a square-shaped universe, bounded by walls, with the _____ at its center.

A. Taiji Hall   B. Son of Heaven
C. Zijin Cheng   D. symbol of humankind

（5）By the time of the 7th through 9th centuries a sophisticated system of one hundred and eight _____ was in place in the capital Chang'an.

A. walled spaces   B. walled wards
C. walled courtyards   D. walled squares

（6）Easy access to a good water supply is _____ to the plan of every Chinese imperial city.

A. characteristic   B. natural
C. essential   D. available

2. Translate the Following Paragraph into Chinese.

The second characteristic of the Chinese imperial city is that outer and inner city walls are pierced by gates. Often, but not always, three gates are found at each outer wall face for a total of twelve. Ideally, gates of opposite city walls are equidistant from the adjoining wall corners. Inner cities of the imperial capital generally have no more than one gate at each side. One central gate always is placed at a south city wall. Most city wall gates are built for entrance or exit by a land route, usually a major urban thoroughfare, but some imperial cities have sluice gates as well.

3. Translate the Following Paragraph into English.

北京皇城规划的指导思想：坚持北京的政治中心、文化中心和世界著名古都的性质；正确处理历史文化名城保护与城市现代化建设的关系；重点搞好旧城保护，最大限度地保护北京历史文化名城；贯彻"以人为本"的思想，使历史文化名城在保护中得以持续发展。

22　Stages and Considerations of Urban Master Planning
（城市总体规划步骤与考虑因素）

教学目标：本单元的主要内容有城市总体规划步骤、城市总体规划考虑因素、城市底图、土地利用图、人口普查、经济研究、社区设施规划以及开放空间。教学重点是城市总体规划步骤，难点是城市总体规划考虑因素。通过本单元的学习，要求理解城市底图、土地利用图、人口普查、经济研究、社区设施规划以及开放空间等，掌握城市总体规划的步骤和城市总体规划需考虑的有关因素。

Urban master plans typically project optimum private and public land use for ten to twenty years. Provision is made for regular review and updating of the plans. Step one in the planning process involves the development of a statement of goals and objectives. Currently there is considerable difference of opinion concerning the degree to which the goals and objectives should be established by planning professionals rather than by citizen groups. *Increasingly planning professionals attempt to utilize techniques whereby citizens can participate with them in setting the initial goals and objectives.* [1]

The second planning stage involves basic studies. A base map is developed, a land use map is developed, economic studies are made, and population projections are made. The third stage of master planning involves the reevaluation of goals and objectives in light of this new and systematic evidence. The fourth stage is the actual writing of the plan. The plan will have chapters or sections dealing with land use, transportation and community facilities. Urban renewal and historic preservation often become specific parts of a master plan.

Implementation of the master plan is carried out through capital improvement programs and by social initiative, both private and public. The police power of the urban government is also used in plan implementation. Urban police power is embodied in zoning ordinances, subdivision regulations, building and housing codes and health and welfare requirements. With each phase of implementation it is desirable to reevaluate the master plan.

A more detailed examination of the elements of the master plan is necessary. The urban base maps identify features such as the street system, railroads, rivers and parks. A major outgrowth of the base maps is a second set of maps showing existing land use facilities in the municipality. A third set of maps shows the topology of the area with special emphasis on subsoil properties and flood plain areas. From a sociobehavioral point of view, cognitive mapping needs to be added at this point. *Physical map information is insufficient without a social map identifying the symbolic meanings and importance of the physical features on the landscape.* [2]

Land use maps increasingly designate subsurface uses, surface uses and air rights or visual uses. As megastructures increase in importance in the urban environment, land use above and below the surface also increases in importance.

Population studies generally have focused on projections indicating anticipated growth in the number of people in urban areas. In recent years, however, some older central cities have experienced decreases in population. Population studies need to be related to cognitive mapping and land use. It is now important that population studies and demographic maps identify the number of people in an area, like a census tract, by variation in a twenty-four-hour period, showing day by day variation and weekday-weekend differences. Some transportation arteries will be "overpopulated" during early morning and late afternoon hours and underutilized at other times. A central business district maybe "overpopulated" during the midday hours of the working week and underutilized during evening hours and weekends. Similarly, residential areas may be "overpopulated" during evening hours and "under-populated" during working hours. The social and economic cost of building urban areas that have over- and under-population at varying times in the day or the week is great. Some patterns of antisocial behavior increase during hours of under-population. Other patterns of antisocial behavior obtain most in hours of overpopulation. *Both types of antisocial behavior could be greatly reduced by establishing master planning goals and implementation techniques to reduce the amount of movement of large populations diurnally between places of work, residence, commerce and recreation.* ③ Megastructures are one alternative in the organization of cities for the integration of socio-physical spaces for high-quality living.

Economic studies traditionally are concerned with the number of jobs that are generated by commerce, industry and government for urban inhabitants. These studies are highly deficient. Most frequently municipalities are adjacent to other municipalities in large urban regions. Furthermore, as many western societies move in a post-industrializing direction and increasingly utilize cybernation, the economics of people and jobs become more complex. Additionally, as birth rates and familism decline, more women are available for positions in the labor force. High-quality urban environments must therefore create satisfactory alternatives for women. Health and longevity also increase; accordingly, more economic provisions are needed for the older populations.

The "community facilities plan" is one of the most critical for achieving a high-quality environment. This is the part of the master plan that deals most with schools and general socialization. It specifies educational and cultural facilities, locating schools, colleges, museums and so forth. Still, this component of the plan deals only implicitly with science, relating it primarily to universities. Master urban planning in the last of the 20th century is done in mind societies where information is largely based on science. A master plan to maximize positive social relationships needs to focus specifically on the location of scientific institutions. Science is too important an element for modern urban life to be implicitly rather than explicitly planned.

Medical facilities are another component of the community facilities plan. In addition to hospitals, these facilities now specify nursing homes, clinics, etc. There is still generally insufficient explicit planning for social welfare beyond health needs. Rehabilitation facilities are

important dimensions of urban life, and they too need to be specified. Government and all public buildings are other components of the community facilities plan. Here there is a detailed specification of the location of fire departments, police auditoriums and related facilities stations, jails, public markets, etc. Churches and related religious facilities identified in this part of the plan are generally given minimal or insufficient attention.

Open spaces and/or parks are a favorite subject of planners. There has been a bias favoring park. There is little hard evidence indicating the extent to which park facilities necessarily improve the quality of the urban environment. ④ Indeed Newman observes that some kinds of large parks are places that harbor antisocial behavior. Moreover, there is little exploration of the development of parklike spaces in megastructures. In the Compact City plan, open park space would be built on top of and into the megastructure building. Similarly, in Soleri's arcology plans, open parklike spaces are integrated into the built environment.

Environment is the last major element of the community facilities plan. Here attention is given to items like water supply, water treatment and water storage. Solid waste facilities and sanitary landfills are specified. Air pollution and flood control are further items for planning attention. Through all of this, there is a limited discussion of recycled environments.

Vocabulary and Expressions：
urban comprehensive planning 城市总体规划
过程
urban base map 城市底图
topology [tə'pɔlədʒi] n. 拓扑学，拓扑关系
demographic [,demə'græfik] map 人口地图
census tract 人口普查区，普查地段
megastructure ['megəstrʌktʃə] n. 大型建筑，
巨型建筑，巨型结构
socio-physical space 社会物质空间，社会物
理空间
cybernation [,saibə'neiʃən] n. 自动控制，计
算机控制

familism ['fæmilizəm] n. 家庭主义(强调家庭和家属感情的社会结构形势)

rehabilitation facility 康复设施

arcology [ɑːˈkɔlədʒi] n. 建筑生态学

diurnally [daiˈɔːnəli] adv. 每日，在白天，白天活动地

Notes：

① 专业规划人士越来越多地尝试利用技术手段使民众可以参与到最初的目标制定中来。

② 仅仅拥有物质空间信息，而没有对景观物理特征的象征意义和重要性进行说明的人文社会地图，是远远不够的。

③ 可以通过建立总体规划目标并将其付诸实施来减少办公、居住、商业、娱乐场所之间的人口的大规模流动，从而减少这两种不良行为。

④ 开放空间和(或)公园是规划师的宠儿，尤其是后者，但是并无有力证据证明公园必然能在某种程度上改善城市环境。

Exercises：

1. Fill in Each of the Blanks with the Correct Answer.

(1) Step one in the master planning process involves the development of a statement of _____.

A. basic studies

B. goals and objectives

C. objectives and approaches

D. component of master plan

(2) The master plan will have chapters or sections dealing with land use, transportation and community facilities, while its specific parts would be urban renewal and _____.

A. reevaluation of destination

B. urban regeneration

C. urban location

D. historic preservation

(3) The police power of the urban government is also used in plan _____, and urban police power is embodied in zoning ordinances, subdivision regulations, building and housing codes and health and welfare requirements.

A. formulation   B. examination

C. implementation   D. establishment

(4) A major outgrowth of the base maps is a second set of maps showing existing land use facilities in the _____.

A. municipality   B. region

C. community   D. society

(5) Master urban planning in the last of the 20th century is done in mind societies where information is largely based on _____.

A. data collection   B. science

C. social collaboration   D. economy

2. Translate the Following Paragraph into Chinese.

The "community facilities plan" is one of the most critical for achieving a high-quality environment. This is the part of the master plan that deals most with schools and general socialization. It specifies educational and cultural facilities, locating schools, colleges, museums and so forth. Still, this component of the plan deals only implicitly with science, relating it primarily to universities. Master urban planning in the last of the 20th century is done in mind societies where information is largely based on science. A master plan to maximize positive social relationships needs to focus specifically on the location of scientific institutions. Science is too important an element for modern urban life to be implicitly rather than explicitly planned.

3. Translate the Following Paragraph into English.

城市总体规划是指城市人民政府依据国民经济和社会发展规划以及当地的自然环境、资源条件、历史情况、现状特点，统筹兼顾、综合部署，为确定城市的规模和发展方向，实现城市的经济和社会发展目标，合理利用城市土地，协调城市空间布局等所作的一定期限内的综合部署和具体安排。城市总体规划是城市规划编制工作的第一阶段，也是城市建设和管理的依据。

23　Urban Transportation
　　　　Planning Phases
（城市交通规划阶段）

教学目标：本单元的主要内容有城市交通特点、城市交通模型系统、城市交通规划阶段、预分析阶段、技术分析阶段以及后分析阶段。教学重点是城市交通特点和城市交通规划阶段，难点是城市交通模型系统。通过本单元的学习，要求熟悉城市交通规划各个阶段：预分析阶段、技术分析阶段以及后分析阶段。理解城市交通模型系统，了解城市交通特点。

23.1　Characteristics of Urban Transportation

Basic to an understanding of transportation are the characteristics of urban travel that many ways define the scope of urban transportation problems. Urban transportation has five characteristics: (1) temporal distribution of urban travel; (2) type of trips made (trip purpose); (3) modal distribution of urban trips; (4) environmental impact of transportation facilities; (5) relationship between land use and transportation.

23.2　Urban Transportation Planning Phases

Transportation planning is undertaken at many levels (from strategic planning to project-level planning) and at different geographic scales in any urban area. Furthermore, the regional studies that became synonymous with urban transportation planning in the 1950s and 1960s have undergone substantial changes over the years. Nevertheless, it is possible to identify a planning process and associated technical analyzes that are commonly considered to be the urban transportation planning process. In this section, we provide a brief overview of this planning process. *We do not present this as a recommended process for urban transportation planning nor do we suggest this as being the way in which urban transportation planning is always undertaken.* [1] Rather, we intend this to be a general framework within which our discussion of specific aspects can proceed.

The urban transportation planning process is viewed here as having three major, interrelated components, namely, the preanalysis phase, the technical analysis phase and the postanalysis phase (Fig. 23-1). A major reason for describing the process in this manner relates to the roles of the various actors (the technical team, the decision makers and the citizens) at various stages of the process. *The activities in the technical analysis phase are conducted almost exclusively by the technical team, whereas the decision makers and the citizens should be involved in the preanalysis and postanalysis phases.* [2]

23.2.1　The Preanalysis Phase

The activities in the preanalysis provide necessary inputs to the technical analysis phase. The preanalysis phase includes problem/issue identification and formulation, the development of study-area goals and objectives, the collection of data concerning the existing transportation and related systems as well as existing travel patterns, and the identification of the alternative solutions to be analyzed. That is, the preanalysis phase has two components. The first concerns defining the current situation and prob-

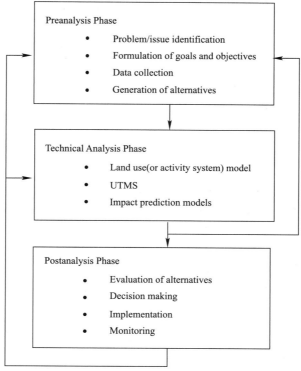

Fig. 23-1   A General Representation of the Urban Transportation Planning Process

lems and specifying the desired characteristics of improvements. *The second aspect of the preanalysis phase includes developing the data to be used in the technical analyzes and formulating the alternative plans and policies to be tested.* [3]

23. 2. 2   The Technical Analysis Phase

The technical analysis phase of the urban transportation planning process is concerned with predicting the impacts of alternative courses of action. In this phase of the planning process, mathematical models are used to predict the transportation and related impacts (consequences) of alternative plans and policies. These impacts include capital and operating costs, energy usage, land requirements, air quality and noise levels and accident rates, in addition to the quantity and quality (e.g., speed) of traffic flow on the transportation network.

The central component of the technical analysis phase is concerned with predicting the quantity and quality of traffic flow on each portion of a specified transportation network. The forecasting techniques used for this purpose are often referred to generically as the Urban Transportation Model System (UTMS). *Important inputs to this model system are the distribution of employment, housing and other activities in the urban area. These distributions are predicted by land use (or activity system) models. The output from the UTMS is the input for a set of impact models that are used to predict the range of impacts described above.* [4]

23. 2. 3   The Postanalysis Phase

The postanalysis phase is concerned with assessing the impacts of the alternative plans and policies, selecting the preferred alternative, implementing the preferred alternative and monitoring the performance of the implemented plans. The monitoring activity emphasizes the continuing nature of the urban transportation planning process.

Vocabulary and Expressions：

Urban Transportation Model System（UTMS）城市交通模型系统

preanalysis phase 预分析阶段

technical analysis phase 技术分析阶段

postanalysis phase 后分析阶段

Notes：

① 我们无意将它推荐作为城市交通规划的过程，也不认为这是城市交通规划的一贯模式。

② 技术分析阶段几乎都是由技术团队主导，而决策者和民众则参与到预分析和后分析环节中。

③ 预分析阶段的第二个方面包括开发技术分析所需的数据，并制定用以进一步评估的方案和政策。

④ 输入城市交通模型系统(UTMS)用以分析的重要信息包括城市地区的职业、住房和其他一些活动的分布情况，这些分布情况可通过土地利用模型（或活动系统模型）进行预测。该系统的分析输出是一系列影响模型的分析输入，用以预测上述各种规划与政策所产生的影响。

Exercises：

1. Fill in Each of the Blanks with the Correct Answer.

（1）Urban transportation planning ＿＿＿＿ the regional studies in the 1950s and 1960s, but has undergone substantial changes over recent years.

A. was associated with

B. was part of

C. was synonymous with

D. was established in

（2）The urban transportation planning process referred to in the text has three major, interrelated components, namely, the pre-analysis phase, ＿＿＿＿and the post-analysis phase.

A. the analysis phase

B. the technical analysis phase

C. the technological analysis phase

D. the mid-term analysis phase

（3）The major reason for distinguishing the three phases of the urban transportation process relates to ＿＿＿＿ at various stages of the process.

A. the roles of the analytical components

B. the roles of the various actors

C. the effects of the technical elements

D. the effects of the different players

（4）In the second phase of the planning process, ＿＿＿＿ are used to predict the transportation and related impacts（consequences）of alternative plans and policies.

A. mathematical models

B. land-use system models

C. impact prediction models

D. dynamic models

（5）The impacts concerned in the second phase of the planning process include capital and operating costs, energy usage, ＿＿＿＿, air quality and noise levels and accident rates, in addition to the quantity and quality of traffic flow on the transportation network.

A. land requirements

B. ecological impact

C. economic benefits

D. sustainable development

（6）The central component of the second phase of the planning process is concerned with predicting the quantity and quality of ＿＿＿＿ on each portion of a specified transportation network.

A. traffic volume   B. traffic noise

C. traffic level   D. traffic flow

（7）In the second aspect of the preanalysis phase, in addition to developing the data to be used in the technical analyzes, ＿＿＿＿ would be formulated.

A. the alternative situation and problems to be estimated

B. the alternative proposals and plans to be verified

C. the alternative plans and policies to be tested

D. the alternative approaches and strategies to be predicted

2. Translate the Following Paragraph into Chinese.

The technical analysis phase of the urban transportation planning process is concerned with predicting the impacts of alternative courses of action. In this phase of the planning process, mathematical models are used to predict the transportation and related impacts (consequences) of alternative plans and policies. These impacts include capital and operating costs, energy usage, land requirements, air quality and noise levels and accident rates, in addition to the quantity and quality (e.g., speed) of traffic flow on the transportation network.

3. Translate the Following Paragraph into English.

城市综合交通规划是指完整的城市交通规划。其综合性贯穿交通规划的全过程，包括：各种历史、现状和未来信息的综合采集、综合处理和综合分析；综合考虑土地使用、交通需求、交通供应三者的相互关系及其预测技术；各种交通运输方式（工具）的综合协调；涉及技术、经济、社会、环境等多种因素的综合可行性研究、评价与决策。

24 Housing Design Criteria and Considerations of Residential Planning

（住房设计准则与住宅规划考虑因素）

教学目标：本单元的主要内容有住房设计准则（实用、坚固、愉悦）、住宅规划需考虑的因素，如客户与场地、社会习惯、家具电器、特殊爱好、客户想法、人体尺度、交通以及特殊用途房间等。教学重点是住房设计准则，难点是住宅规划需考虑的因素。通过本单元的学习，要求理解住房设计准则：实用、坚固、愉悦，掌握对住宅规划需考虑的主要因素的深层次思考。

24. 1　Housing Design Criteria

Housing is living quarters for human beings. The basic function of housing is to provide shelter from the elements, but people today require much more than this of their housing. A family moving into a new neighborhood will want to know if the available housing meets its standards of safety, health and comfort. A family may also ask how near the housing is to churches, schools, stores, the library, the movie theater and the community center.

Good housing means a satisfactory community, as well as proper shelter. Residential neighborhoods should have a maximum of quiet, privacy, cleanliness and safety. They should be served by hospitals, schools, police, fire, sanitation and street departments. Parks and social centers weld a group of individual homes into a community. These conditions create good standard housing.

Any discussion about housing design must involve concepts of beauty and how beauty is evaluated. Because beauty is thought of as an elusive idea we are reluctant to acknowledge it. Some aspects of beauty and design can only be dealt with subjectively, however, most are easily understood and can be reviewed objectively. *Beauty is not abstract, it is real. Beauty is not necessary in the eye of the beholder; it is largely objective and quantifiable. There are three criteria which Sir Henry Wotton paraphrased in the 17th century, from the first century B. C. writings of Vitruvius, that are still applicable today and are the basic measuring blocks of housing design. These criteria are: commodity, firmness and delight.* [1]

Commodity poses the question, is the building suitable for its purpose and does it look like what it is? *We are talking about housing, so ask yourself, does it look like a Canadian house and not like a Swiss Chalet or a Jacobean Manor or a California Cinderella Bungalow?* [2] Are the materials appropriate to the location and are they arranged so that they respect the existing surroundings? The honest use of building materials does not mean building houses only with wood and stone; glass and aluminum may be more appropriate, indeed, may be more natural in some circumstances. Suitability for purpose also affects the floor plan and this is particularly critical in small house. Does the circulation work well? Is the house zoned to provide both living and quiet areas? Is there adequate space for cooking, cleaning and all the activities connected with bringing up a family? Firmness deals with the adequacy of the construction. Are good materials used and are they used technically correctly and with good workmanship? Do the walls and the roof give suitable protection from the weather? Is there no undue maintenance? A building that is falling to bit because of its poor construction cannot be an object of beauty. Delight is a wonderfully chosen word; it means to give pleasure to the beholder. Again, we are not talking about abstract concepts but about quantifiable realities. *Delight has to do with scale, proportion, harmony, rhythm and unity—with what the building, in its setting, looks like.* [3]

Scale concerns the size relationship between people and buildings or places. When we look at a building we want clues about how to interpret its size. Our eyes seek out things we know. We are familiar with the size of a brick and the dimensions of an ordinary door and we are comfortable when the sizes turn out to be what we anticipate. Another aspect of scale involves our feeling of importance as individuals. If we appear large in relation to our surroundings we feel that we can cope with the elements of our environment.

Proportion is the comparable relationship between the sizes of number of things, or between the sizes of parts of one thing. A window has a size relationship of width to height. A room has a size relationship of width to length to height. The side of a house is related in size to the sizes of the windows or doors which it may contain. At a large scale, proportion deals with the size of an open space, such as a play yard, compared with the height of the buildings around it. Some proportions are more pleasing to the eye than others. And there have been many attempts to formulate rules that will give satisfying size relationships. The proportions of the human body are often a source of intense beauty.

Harmony comes from putting things together in such a way that the individual parts of the building make a consistent and orderly statement. It is inappropriate to mix styles within one building or to mix materials at random. For example, if the principal exterior material is brick, harmony is lost if there are also large areas of stone, or stone and wood. A group of buildings should present a consistency in their shapes, their exterior materials, the colors used and so on. This does not imply rigid uniformity but suggests that variation should be kept within a limited range of agreeable alternatives.

Rhythm is a fundamental part of our life processes. There is a rhythm to our heart beat, to the way we breathe. Changes in rhythm between sleeping and waking are a fundamental factor in our lives. Therefore,

rhythm appeals to the deepest roots of our being and provokes a strong emotional response. In music the simplest form of rhythm is hand clapping and in dance it is being in step. Hand clapping and being in step have to do with time intervals. The same is true in architecture; the distance between the elements creates the pattern.

Unity means oneness. Obviously, there will be many parts in a design; but we should be able to recognize the functions of the parts and the rightness of how they have been put together to make a whole. A building must relate to what the observer already knows. In order to do this the architect must have a clear concept of his project in human and social terms. In developing the design he must emphasize everything that express this concept and eliminate everything that detracts from it.

Good design further implies the need to take into account what is happening in the world around us. This applies not just to the individual house but to the design of the whole community. Particularly important nowadays is the need to understand the relationship between housing and energy. Suburban sprawl leads directly to high energy consumption. The way we lay out suburbia precludes an efficient public transportation system; we are burning up the future of our country driving to the corner store and the hockey rink. The ideal of the detached house with two cars in the garage, isolated on as large as a lot as possible, is inconsistent with today's energy concerns. Commodity, firmness and delight may be old, but they still apply today. Meeting these criteria is not a limiting experience; a Gothic Cathedral is built according to them. Given the materials we now have available and pressing concerns with issues of conservation, energy and transportation, we know our buildings must be different from historic examples. The architect can select and arrange materials to form beautiful buildings with greater freedom than ever before.

24.2   Considerations of Residential Planning

(1) The client and the site: The client and the site together determine the house. Do not try to design a house for someone until all the information regarding the site has been assembled. A house that works well when faced north might be unsatisfactory when faced west. At all stages of planning, the relationships of the three basic areas of the house (working, living and sleeping) to the site must be kept in mind. The client's needs and desires must be the starting point in planning, as the house will be for him. Some of the factors involved are simple and need no explanation, while others are more complex and will be discussed. Some obvious points are: (a) The size of the family determines the size of the house; (b) The distribution by age and sex determines the number of bedrooms; (c) The number of automobiles determines the size of the garage or *carport*[④] (Fig. 24-1); (d) The amount of money available for building will affect the quality and size of the entire project (Cost estimating is a field in itself; the student should consider costs in general terms only).

Fig. 24-1   Carport

(2) Social habits: Some families live very formally and have no use for a family, rumpus room or same room. Others will need such a special-purpose room to suit their informal way of life. Durable, easily cleaned walls and floors are necessary if the family likes to entertain and give parties often. An open plan, in which the living and working areas flow together, lends itself to informal entertaining. A more rigid plan, in which these areas are closed off from each other, works better for formal living. Extra bedrooms and baths will be needed if the family often entertains overnight guests.

(3) Furniture and appliances already owned by the family: Standard-size furniture will not cause much trouble, but oversize pieces can create problems. Be sure they will fit into the house.

(4) Special interests and hobbies: Books, collections and hobbies all require space. Often, special storage or rooms must be added for them. For instance, hobbies like wood-working and photography require special rooms; stamp collecting generally does not.

（5）The client's ideas: His ideas for the number of bathrooms, architectural style, colors and so on, should be worked into the house as far as possible. If they cannot be used, show him why not.

（6）Human scale: Houses are built for people. Therefore, all parts of a house should be built to the scale of the owners and the things they use. A 1.524m ceiling, a 2.323m^2 bedroom, a 1.524m high table—all of these are obviously ridiculous and would not work; however, other errors are not so easy to see. For instance, a 762mm high built-in food bar is just right, while 609.6mm would be too low to use with the average chair. Before starting, check *Architectural Graphic Standards* (GB/T 50104—2010)[5] for the sizes of furniture, appliances, standard working heights and so forth.

The first step in the development of the floor plan is to determine the approximate area requirements of the client, and to separate them into the three basic areas of a house: sleeping area, living area and working area. The requirements for these areas can be taken from the client's checklist. The sleeping area includes the bedrooms, den, bath, storage and halls in that part of the house. The living area includes the living room, family room and dining room. The working area includes kitchen, utility room, pantry and possible bath. Where rooms serve two purposes, such as working and living, be sure to keep the areas close together. Room for the furnace or air-conditioning unit must be provided in the area in which it is to be placed. Most manufacturers produce unit air conditioners for roof mounting which contain a furnace, cooling coils, motor and condenser. Such a unit completely eliminates the need for a furnace room. However, the units are fairly large and might be unacceptable to some people because of their appearance.

The sleeping, living and working areas are represented by "balloons" in the first sketches. The first sketches should be drawn on the site plan, usually a plan of the lot. In this way the basic areas of the house can be properly to take advantage of the weather, the view, and access to the property. The garage walks and driveways, service areas, outdoor living, and major landscaping can be planned before any detailed floor planning is attempted. This procedure will avoid many problems in the final planning of the individual working drawings. There are many general arrangements of the rooms in a house which will satisfy different people. Try to plan for efficient use of space and ease of traffic flow. A particular client may want changes made from the most efficient arrangement and may resist all logical argument. Remember, part of the designer's job is to satisfy the client.

（7）Traffic: The problem of traffic flow should be considered first—that is, the most-used areas of the house should be kept fairly close together to avoid unnecessary steps. As the housewife is in the house much more than the rest of the family and is responsible for its upkeep, it stands to reason that "traffic" refers mainly to her traffic.

（8）Specific rooms: The basic requirements of the main rooms of the house, starting in a logical order the front door, are listed here:

（a）Entry: *An entry is a great convenience*[6]. Though nor required in all houses, the use of an entry is generally considered preferable to the alternative of having the front door open directly into the living area. The entry should, when possible, provide access to the living, working and sleeping areas of the house. It has no required size or shape as long as it works. It may be completely walled or set off by room dividers, or it may be an extension of the living, family or dining room.

（b）Living room: The living room should be fairly close to the front door but should not be a passageway to other parts of the house. It should be near the kitchen or dining room for ease of entertaining and hospitality. Isolate it from the bedrooms for quiet and privacy in the bedrooms. An average size for a living room is about 27.871m^2.

（c）Kitchen: This is the most important room in the house from the standpoint of the housewife. It should be near the front door for admitting front-door visitors; adjacent to the utility room for washing, storage and so on; and near the back door for facilitating traffic to the service yard and garage. An average size for a kitchen is 11. 148m². The draftsman must have definite information about the desired type of kitchen before he starts the preliminary floor plan. Many shapes and sizes of kitchen are possible, and housewives have firm ideas about what type they want. The location of major fixtures and appliances is of first importance. The sink, stove, oven, refrigerator and freezer must be so located as to eliminate extra steps. A short triangular path between sink, stove and refrigerator, the most-used fixtures, is considered desirable. The freezer and oven are not used as often, and therefore may be placed out of the mainstream. The location and type of doors greatly affect the efficiency of the kitchen. The width of the aisle between counters is also important; for instance, too wide an aisle in a corridor kitchen would eliminate the principle advantage of this arrangement. A student interested in kitchen planning can find a great deal of information in manufacturers' brochures, textbooks and magazines.

(d) Dining room: A dining room is not always needed when a dinette, family room or similar room is included in the plan. But when it is required, it should be adjacent to the kitchen and close to the living room for serving and hospitality. An average size for a dining room is 11. 148m².

(e) Family room, all-purpose room, den, rumpus room and so on: Many contemporary homes have such rooms. Some of them are used mainly for dining and may be extensions of the kitchen. Others are entirely separate and are used for television viewing, hobbies, entertaining and sometimes sleeping. In any case a room of this type should be near the kitchen and patio or terrace. Depending on its main use, it should be near the living room or the sleeping area of the house. An average size for an all-purpose room is 22. 297m².

(f) Utility room: It should be adjacent to the kitchen and back door and should provide easy access to the service yard and garage. An average size for a utility room is 5. 574m².

(g) Bedrooms: Bedrooms should be isolated from the noisy areas of the house, adjacent to the bathroom, and preferably arranged around a storage hall with a bathroom. An average size for a bedroom is 12. 077m².

(h) Bathroom: At least one bathroom should be in the bedroom wing of the house. If there is more than one bath, the other could be near the back door and utility area, or adjoining the master bedroom. An average size for a bathroom is 4. 645m².

(i) Halls: When space and expense must be considered, halls should be kept small. One hall should be in the bedroom wing and have access to the bathroom and linen closets. Do not try to omit halls entirely to save space; they are necessary in an efficient planned house. The recommended minimum width for a hall is 1. 067m.

(j) Basement: When a basement is included, it can be placed under any desired part of the house. The entrance may be either outside or inside. Avoid a basement entrance from the bedroom wing or living room. Place heating and ventilation equipment, water heaters and water softeners in the basement. A large basement is useful as a rumpus room or storage area. A basement may be any desired size.

(k) Heater room: When no basement is provided and a central heating and/or cooling plant is called for, the equipment may be put in the garage or in a special heater room. The room should be designed around the particular pieces of equipment desired. Access must be provided for installation and maintenance of the equipment. It must be located at a point where it can easily be hooked up to ducts and pipes serving the building. An average size for a heater room housing a gas-fired unit is 1. 208m².

(l) Storage space: Space must be provided in each of the three main areas of the house for the things that people own. Generally, too much storage is better than too little. In the living area of the house, provide a coat closet and space for books, hi-fi equipment, television, games, extra tables and chairs and so on. The linen closets in the sleeping area of the house must be large enough for linens, extra blankets and bathroom supplies. The closets should be about 457.20mm deep. Bedrooms and dens should have adequate wardrobe or closet space for closing. The largest amount of storage space is needed in the working area. Various types of storage must be allowed for: dry, canned, frozen and fresh foods; dishes; silverware; food utensils; linens; cleaning equipment and staple items. For more complete information, check *Time-Saver Standards*,[⑦] or ***Architectural Graphic Standards*** (GB/T 50104—2010). The designer should remember that storage requirements will vary greatly with clients. Be sure to get the client's ideas about storage.

(m) Mud room: In cold or wet climates, such a room is placed near the building entrance to take care of the problems of changing and storing wet clothes. Floor and wall surfaces should be durable and easily cleaned.

(n) Garage or carport: Requirements for car storage vary considerably from place to place. In all cases it is necessary to consider size, location, access and other uses for the space.

(o) Miscellaneous: Many other considerations concerning the functioning of the entire project affect the site and floor plans. Provision must be made for installation, access and maintenance of local chutes, fuel oil tanks, *LPG tanks*,[⑧] utility meters, wells, swimming pools, *septic tanks*,[⑨] hot water heaters and so on.

If the house has more than one level, draw the proper stair sections to check for clearance in stairwell, access to rooms and space required for stairs and halls. When the relationships between the rooms become definitely established, check the plan for unnecessary "jogs" and setbacks in the wall lines. Unless needed for functional reasons, corners cost money but contribute little to the livability of the house. Proper orientation of the building to the sun and weather must be considered at all times as the floor plan is developed.

Vocabulary and Expressions：

elusive [i'lu:siv] adj. 难懂的，难捉摸的

hand clapping 鼓掌，拍手，掌声响起

single-family split-level house 独户错层式住宅

Gothic house 哥特式建筑

Byzantine（英［bai'zæntain］，美［'bizənti：n]）architecture 拜占庭式建筑

hockey rink 冰球场

family or rumpus room 娱乐室

pantry ['pæntri] n. 餐具室，食品室或备餐室

terrace ['terəs] n. 露台

den [den] n. 书房，工作室

patio ['pætiəu] n. 天井，院子

dinette [dai'net] n. 厨房旁小餐室，小吃饭间

Architectural Graphic Standards（***GB/T 50104— 2010***）《建筑制图标准》GB/T 50104—2010

mud room（农舍入口的）小屋，泥鞋室

jog [dʒɔg] vt. 慢跑，蹒跚行进；vi. 凹进，凸出，慢跑，蹒跚行进；n. 凹进，凸出，慢跑

Notes：

① 美不是抽象的，它是真实的。在观众眼里，美并不是必需的。美在很大程度上是客观的、可量化的。这里有来自公元前1世纪维特鲁威作品的后来在17世纪被亨利·沃顿爵士重新诠释过的三项原则，这三项原则今天仍然适用，是衡量建筑设计好坏的三大基本模块。这三项原则是：实用、坚固、愉悦，也即适用、安全、美观。Sir Henry Wotton 亨利·沃顿爵士 architectural theorist of United Kingdom（UK），著书《建筑学要素》（The Elements of Architecture）。Vitruvius 维特鲁威，the greatest architect of ancient Rome，著书《建筑十书》（The Ten Books on Architecture）。

② 我们正在谈论房子的问题，可以问问你自己（或可以扪心自问），它看起来像加拿大的房子而不像瑞士（阿尔卑斯山）小（木）屋或詹姆斯一世（时期风格）的庄园或加州

（普通）平房吗?

③ 住宅的舒适愉悦性包括合适的比例、均衡、调和、节奏韵律以及整体和谐的效果。这些才是建筑环境所需要的。

④ carport 汽车棚。A covered automobile shelter associated with a separate dwelling. It has one or more sides open to the weather. 有顶棚,四周至少有一面是敞开的。

⑤《建筑制图标准》GB/T 50104—2010。

⑥ 出入口是一个方便进出的空间(或出入口是一个非常便利的设施)。

⑦ Time-Saver Standards for……标准速查手册。如 Time-Saver Standards for Architectural Design Data 建筑设计资料速查手册。

⑧ LPG tank(Liquefied Petroleum Gas tank)液化(石油)气罐。

⑨ septic tank(cesspit)化粪池。

Exercises:

1. Answer the Following Questions in English to Help Comprehension and Appreciation.

(1) What are the main criteria which Sir Henry Wotton paraphrased in the 17th century?

(2) How many aspects which are included in the concept of "delight"?

(3) What's the important content of proportion?

(4) What does the "harmony" mean for the architecture of housing?

(5) Why did the author say delight has to do with scale, proportion, harmony, rhythm and unity?

(6) Give the English explanation of the detached house.

(7) What does the good housing design imply?

(8) Would you please sum up main elements of good housing design?

(9) What are clients' needs and desires for housing living?

(10) How about the social habits to be considered?

(11) How to consider the point of human scale in residential planning?

(12) What are the three basic areas in residential planning?

(13) Are there many general arrangements of rooms in a house?

(14) What are the family room, all-purpose room, den and rumpus room?

(15) What is the function of the mud room in a house?

(16) Please think of all the main points of residential planning.

2. Translate the Following Paragraph into Chinese.

The first step in the development of the floor plan is to determine the approximate area requirements of the client, and to separate them into the three basic areas of a house: sleeping area, living area and working area. The requirements for these areas can be taken from the client's checklist. The sleeping area includes the bedrooms, den, bath, storage and halls in that part of the house. The living area includes the living room, family room and dining room. The working area includes kitchen, utility room, pantry and possible bath. Where rooms serve two purposes, such as working and living, be sure to keep the areas close together. Room for the furnace or air-conditioning unit must be provided in the area in which it is to be placed. Most manufacturers produce unit air conditioners for roof mounting which contain a furnace, cooling coils, motor and condenser. Such a unit completely eliminates the need for a furnace room. However, the units are fairly large and might be unacceptable to some people because of their appearance.

3. Translate the Following Paragraph into English.

房屋设计是为人类生活与生产服务的各种民用与工业房屋的综合性设计。根据选用的材料,配合周围环境,在安全、适用、经济和美观之间寻求合理的平衡。房屋设计的产品为建筑、结构、设备各专业的图纸与说明书及其概算,可作为房屋施工的依据。

25 Bauhaus—"Art and Techniques, a New Unity"

（包豪斯——"艺术与技术的新统一"）

教学目标：本单元的主要内容有包豪斯学校发展历程、包豪斯教学理念、包豪斯设计理论以及包豪斯的影响。教学重点是包豪斯学校发展历程，难点是包豪斯教学理念和包豪斯设计理论。通过本单元的学习，要求理解包豪斯设计理论，掌握包豪斯教学理念，了解包豪斯学校发展历程和包豪斯的影响。

The *Bauhaus*[①] was a school of design, building and craftsmanship founded by *Walter Gropius*[②] in Weimar in 1919. It was transferred to Dessau in 1925, where it continued until 1932, and then transferred to Berlin, ultimately closing in 1933. The ideas and teaching of Bauhaus have exercised a profound influence throughout the world.

Before the First World War the Belgian architect *Henry Van de Velde*[③] had been director of *the Grossherzogliche Saechsische Kunstgewerbeschule and the Grossherzogliche Saechsische Hochschule fuer Bildende Kunst*[④] at Weimar, and he had recommended to the Grand Duke of Saxe-Weimar that Walter Gropius should be his successor. The Grand Duke summoned Gropius for an interview in 1913, and Gropius asked for and was given full powers to reorganize the schools; when he took up his appointment in 1919, he united the two schools under the name of *Das Staatliche Bauhaus Weimar*.[⑤] This was of profound significance because it made clear at the outset that one of the main purposes of the new school was to unite art and craft which had for too long been divorced from each other. Gropius contended that the artist or architect should also be a craftsman, that he should have experience of working in various materials so that he knew their qualities and that he should at the same time study theories of form and design. The traditional distinction between artist and craftsman should, Gropius thought, be eliminated. He also believed that a building should be the result of collective effort, and that each artist-craftsman should contribute his part with a full awareness of its purpose in relation to the whole building. Gropius was therefore an advocate of team-work in the creation of a building and in the production of furniture, pottery and all the various architectural arts.

The teaching thus comprehended industrial production. *Gropius was not opposed, as was William Morris, to the increasing use of machinery in the production of well-designed objects, but he believed that the machine should be made absolutely subservient to the will of the creative designer.*[⑥] This part of Gropius's teaching has perhaps been most difficult for many people to understand. Many critics have asked why it is necessary for students to master a craft in a material and yet acquiesce in industrial production. But Gropius regarded machinery merely as an elaboration of the hand tool of the craftsman, and thought it was necessary to know the nature of the material and all its potentialities before the tool or machine could be used to the best advantage.

There is obviously a correlation of team-work in building and the necessary division of labor in industrial production, but the best results are likely to be obtained in both if the members of the team not only master their own particular part but grasp its relation to the complete building or industrial product. By thus using the machine to the best advantage the training at the Bauhaus was directed not to works of hand craftsmanship but to the creation of type forms which could serve as models for mass production. And in the creation of this type-form the artist himself produces the prototype, that is if it is a teapot he makes this in the clay with his own hands as the model for mass production; he is no longer merely the drawing-board designer, but the designer-craftsman.

The curriculum of training consisted of two parallel courses of instruction, one devoted to the study of materials and craft (Werklehre) and the other to the theories of form and design (Formlehre).[⑦] In the early years of the school it was necessary for the student to be taught by two masters, one in each section, an artist and a craftsman, because of difficulty at that time of finding teachers who were sufficiently masters of both. These two teachers worked in close collaboration. Instruction at the school began with a preliminary course of six months, during which period the student worked with various materials—stone, wood, metal, clay, glass, pigments and textile—while he received elementary instruction in the theory of form. The purpose of working and experimenting with materials was to discover with which particular material the student had naturally the most creative aptitude, for it was an essential purpose to bring out the latent creative faculties of the individual. It might be that one student had a strong feeling for wood, another for the harder materials, stone and metal, another for textile, another for pigments and color. He was instructed in the use of tools and later in the use of machines that in industry have supplanted these tools. In the school devoted to form and design, instruction was given in the study and representation of natural form, in geometry and principles of building construction, in composition and the theories of volume, color and design.

　　Gropius was fortunate in gathering together some very able teachers, many of whom afterwards became famous in their various spheres. Among the first was Johannes Itten, who joined the school in 1919 and whom Gropius had first met a year earlier teaching in a private school in Vienna. Gropius was impressed with Itten's methods of education and invited him to the Bauhaus to direct the preliminary course. Itten's teaching included the study of the physical character of natural materials by representation and experiment, for it was contended that representing a material intelligently was one way of appreciating its structure. And in working in a particular material the student must not only develop a feeling for it in all its aspects, but he must appreciate its relation to other materials so as to be aware of its qualities by comparison and contrast. Other teachers at the Bauhaus at Weimar were Lyonel Feininger, printer and graphic artist, and Gerhard Marcks, sculptor and potter, both of whom joined in 1919; Georg Muche, painter, weaver and architect joined in 1920; Paul Klee, painter, graphic artist and

writer, and Oskar Schlemmer, painter and stage designer, both joined in 1921; Wassily Kandinsky, painter and graphic artist, who joined in 1922; and Laszlo Moholy-Nagy, painter, theatrical designer, photographer and typographer, who joined in 1923.

　　In 1923, at the request of the Thuringian Legislative Assembly, the Bauhaus held an exhibition of its work which was to serve as a report on the four years of the life of the Bauhaus. Gropius felt that this was a bit premature; he would have preferred to wait until more mature results could be presented. The theme of the exhibition was "Art and Techniques, a New Unity", and included in the exhibition were designs in various materials, various products of the different workshops, examples of theoretical studies, and a one-family house called "Am Horn" which was built and furnished by the Bauhaus workshops. [8]

　　This house was planned as a large square with several small rooms arranged round a central larger one; it was enthusiastically acclaimed by many critics, among them *Dr. E. Redslob, the National Art Director of Germany*, [9] who praised its organic unity.

　　In spite of the progress and success of the Bauhaus, it met with much local opposition from the more conservative members of the community, while the whole enterprise was associated with Socialism in the minds of many because it happened to be established at a time when there was a Socialist regime. It also met with considerable hostility from the Thuringian Government, which more or less forced Gropius to a decision at the end of 1924 to close the school. Both teachers and students wholeheartedly supported Gropius; the Director and masters notified the Government of Thuringia on 26 December of their decision to close the institution, created by them, on the expiration of their contracts on 1 April 1925.

　　Various cities discussed the possibility of transplanting the Bauhaus, among them Frankfurt, Hagen, Mannheim, Darmstadt and Dessau. The Mayor of Dessau succeeded in securing the transfer of the Bauhaus to his town. He appropriated seven houses for the use of the school while a new building was being erected. This building, designed by Gropius in response to the request of the City Council, was begun in the autumn of 1925 and completed in December 1926. It consisted of three principal wings, a school of design occupying one, workshops another and a students' hostel a third. The first two were linked by a bridge over a roadway, and in this bridge were administrative rooms, club rooms and a private atelier for Professor Gropius. The students' hostel was a 6-storey building consisting of twenty-eight studio-dormitory rooms. The building was constructed partly of reinforced concrete. In the workshops' wing reinforced concrete floor slabs and supporting mushroom posts were employed with the supports set well back to allow a large uninterrupted glass screen on the facade extending for three storeys. This was probably the first time so ambitious a use of glass screen was employed in an industrial building, and it helped to lead the way to similar constructions throughout Europe and America.

With the re-establishment of the Bauhaus at Dessau, the opportunity was taken to revise the curriculum. The earlier method of joint instruction by two masters, an artist and a craftsman, was abandoned and was supplanted by that of one master who was trained as both. This was becoming increasingly possible because several former Bauhaus students were now appointed masters: Josef Albers, Herbert Bayer, Marcel Breuer, Hinnerk Scheper and Joost Schmidt. Seven of the masters who had been with the Bauhaus at Weimar continued at Dessau. Gerhard Marcks left because there were not sufficient funds to install his pottery workshop at Dessau. Johannes Itten had left in the spring of 1923 owing to differences of opinion on the conduct of the preliminary course, and his work was continued by Moholy-Nagy and Josef Albers, who jointly broadened its scope. In revising the course at Dessau the opportunity was also taken to reaffirm the principles which guided the Bauhaus system of education: these

could be summarized as training in design, techniques and craftsmanship for all kinds of creative work, especially building; the execution of experimental work, especially building and interior decoration; the development of models or type-forms for industrial production and the sale of such models to industry. As a general doctrine the Bauhaus sought to establish the common citizenship of all forms of creative work and their interdependence on each other.

The Bauhaus continued at Dessau under the direction of Walter Gropius until early in 1928, when he resigned because he wished to devote himself more freely to his creative work without being restricted by official duties; on his recommendation Hannes Meyer, the Swiss architect, who had been head of the department of architecture, became director. Meyer resigned in June 1930 as the result of differences with the municipal authority, who then tried to persuade Gropius to take over again. Instead, Gropius recommended the appointment of *Ludwing Mies van der Rohe*®, who accepted the position. In October 1932, after the National Socialist party had taken over the Government of Anhalt, the Bauhaus moved to Berlin; in April 1933 it was closed by the National Socialists. From April 1933 the building at Dessau was used for the training of political leaders.

Although the school was closed, its teaching and methods were by no means dead, and they continued to exercise a wide influence throughout the world. Indeed, it may be said that its influence has been strongest since it ceased to exist, probably because it takes time for such ideas to spread. Many art schools in Europe and America have adopted in part its methods of teaching, especially as many of its masters and students have taken positions in art schools and institutes throughout Europe and America. For example, Moholy-Nagy became director of the New Bauhaus—now the Institute of Design—at Chicago, where Bauhaus methods were employed. They have also been introduced partially at the school of architecture at Harvard University, the Laboratory School of Industrial

Design in New York and in the Southern California School of Design. It would be a mistake to think that the ideas that prompted Bauhaus training are universally accepted, but it is doubtful whether any method of art teaching of the century has had quite the same impact.

Vocabulary and Expressions：

Weimar ['waimɑ:] n. 魏玛市（德国城市）

Dessau ['disɔ:] n. 德绍（德国城市）

Moholy-Nagy ['mɔhəli'neigi] n. 莫霍利·纳吉（建筑师）

the Grand Duke of Saxe-Weimar 萨克森-魏玛大公爵

type-form 形制，造型，型制

prototype ['prəutətaip] n. 创作原型，样品

artist-craftsman 艺术家兼手艺人

designer-craftsman 设计师兼手艺人（或设计师兼工匠）

William Morris ['wiliəm 'mɔris] n. 威廉·莫里斯（建筑师）

Johannes Itten [dʒəu'hænis 'itən] n. 约翰内斯·伊顿（建筑师）

Lyonel Feininger ['liənəl 'fainiŋə] n. 利奥尼·费宁格（建筑师）

Gerhard Marcks ['gə:həd ma:ks] n. 格哈德·玛克斯（建筑师）

Georg Muche [dʒɔ:dʒ mʌk] n. 乔治·蒙克（建筑师）

Paul Klee [pɔ:l kli:] n. 保罗·克利（建筑师）

Oskar Schlemmer ['ɔskə 'ʃlemə] n. 奥斯卡·史雷梅尔（建筑师）

Wassily Kandinsky ['wɔsili 'kændinski] n. 瓦西里·康定斯基（建筑师）

Laszlo Moholy-Nagy ['læzlou 'mɔhəli 'neigi] n. 拉兹洛·莫霍利·纳吉（建筑师）

Frankfurt ['fræŋkfət] n. 法兰克福（德国城市）

Hagen ['ha:gən] n. 哈根（德国城市）

Mannheim ['mænhaim] n. 曼海姆（德国城市）

Darmstadt ['dɑ:mstæt] n. 达姆施塔特（德国城市）

Josef Albers ['dʒəusif 'ælbəs] n. 约瑟夫·阿伯斯（建筑师）

Herbert Bayer ['hə:bət 'beijə] n. 赫伯特·拜耶（建筑师）

Marcel Breuer ['ma:səl 'bru:ə] n. 马塞尔·布鲁尔（建筑师）

Hinnerk Scheper ['hinnək 'ʃi:pə] n. 辛涅克·谢柏（建筑师）

Joost Schmidt [dʒu:st ʃmit] n. 乔斯特·史密特（建筑师）

Hannes Meyer [hæns 'meijə] n. 汉斯·迈耶（建筑师）

atelier [ə'teliei] n. 工作室，画室

Laboratory School of Industrial Design in New York 纽约工业设计实验学校

Southern California School of Design 南加州设计学校

Notes：

① "包豪斯"是德语 Bauhaus 的音译，把德语 Hausbau（房屋建造）颠倒一下变成 Bauhaus（造房子），是指格罗皮乌斯 1919 年在德国魏玛创立的德国古典现代主义中最为著名的一个艺术和设计流派，该派在 1919～1933 年产生的创作（建筑、工艺设计）对世界建筑艺术和工艺设计的发展产生了非常重大的影响。继 1996 年魏玛和德绍的包豪斯建筑被列入世界文化遗产名录之后，2004 年教科文组织又将以色列特拉维夫市中心的约 4 千多栋包豪斯成片建筑列入名录。一个现代建筑设计流派的建筑两次被列入名录是极为罕见的。

② Walter Gropius ['wɔ:ltə 'grəupiəs] 瓦尔特·格罗皮乌斯，1883 年 5 月 18 日生于德国柏林，是德国现代建筑师和建筑教育家，现代主义建筑学派的倡导人和奠基人之一，公立包豪斯学校的创办人。积极提倡建筑设计与工艺的统一，艺术与技术的结合，讲究功能、技术和经济效益；设计讲究充分的采光和通风（通透的玻璃幕墙）；主张按空间的用途、性质和相互关系来合理组织和布局，按人的生理要求、人体尺度来确定空间的最大极限等；极力主张用机械化大量生产建筑构件和预制装配的建筑方法。

③ Henry Van de Velde 亨利·凡·德·威尔德（比利时建筑师）。

④（德语）萨克森高等工艺美术学校（Saxe Vocational School of Arts and Crafts）和萨克森

高等建筑艺术学院（Saxe College of Architectural Art）。

⑤ （德语）魏玛国立建筑学校（the State Bauhaus，国立包豪斯学校）。

⑥ 格罗皮乌斯不像威廉·莫里斯那样反对增加优秀设计作品生产中的机器使用，但是他认为在具有创造性的设计师的理念中，机器的使用要放在绝对从属的地位。未省略原句为：Gropius is not opposed, as William Morris is opposed …，在 as 引导的从句中省去 opposed。

⑦ 训练课程包括两类并行的教学课程，一类专攻材料和工艺研究（德语，作坊大师），另一类学习形制和设计理论（德语，形制大师）。

⑧ 1923 年在图林根立法会的要求下，包豪斯举办了一次作品展，其作品作为一份关于包豪斯四年发展的纪录。格罗皮乌斯认为这一展览有点为时过早了；他更愿等到多一些成熟作品再拿出来展览。展览主题是"艺术与技术的新统一"。展览包括以各种材料所作的设计，各车间的各种不同产品，理论学习中的实例，还有一栋由包豪斯创作室建造装修的称作"号角"的独立式住宅。这栋住宅在平面上设计成一个大正方形，由周边几个小房间围绕着中央的一个大房间。结果大受评论家们的赞扬，德国民族艺术总监德斯洛布教授高度称赞此设计的有机统一。

⑨ 德国民族艺术总监德斯洛布教授。

⑩ Ludwing Mies van der Rohe 密斯·凡·德·罗（1886～1969，Ludwin 路德维格，一般不译）是 20 世纪最著名的四大建筑师之一，被誉为"玻璃幕墙的缔造者"。名言：Less is more. 少就是多。

Exercises：

1. Answer the Following Questions in English to Help Comprehension and Appreciation.

（1）What does the Bauhaus mean, a school or institute?

（2）When the Bauhaus was founded?

（3）Why part of Gropius's teaching has been most difficult for many people at the beginning of the 20th century?

（4）What are the two main parallel courses for the curriculum of training?

（5）What are advantages of Bauhaus's education?

（6）What's the theme of the exhibition of Bauhaus in 1923?

（7）How to revise and reaffirm the principles which guided the Bauhaus system of education?

（8）Who is the American director of the New Bauhaus—now the Institute of Design at Chicago?

2. Translate the Following Paragraph into Chinese.

There is obviously a correlation of teamwork in building and the necessary division of labor in industrial production, but the best results are likely to be obtained in both if the members of the team not only master their own particular part but grasp its relation to the complete building or industrial product. By thus using the machine to the best advantage the training at the Bauhaus was directed not to works of hand craftsmanship but to the creation of type forms which could serve as models for mass production. And in the creation of this type-form the artist himself produces the prototype, that is if it is a teapot he makes this in the clay with his own hands as the model for mass production; he is no longer merely the drawing-board designer, but the designer-craftsman.

3. Translate the Following Paragraph into English.

在设计理论上，包豪斯提出了三个基本观点：①艺术与技术的新统一；②设计的目的是人而不是产品；③设计必须遵循自然与客观的法则来进行。这些观点对于工业设计的发展起到了积极的作用，使现代设计逐步由理想主义走向现实主义，即用理性的、科学的思想来代替艺术上的自我表现和浪漫主义。

26　Hood Museum of Art
（护德艺术博物馆）

教学目标：本单元的主要内容有护德艺术博物馆的设计特色、森特布鲁克建筑师事务所、莱斯罗普画廊以及达特茅斯绿地。教学重点和难点均为护德艺术博物馆的设计特色。通过本单元的学习，要求熟悉并理解护德艺术博物馆的设计特色，了解森特布鲁克建筑师事务所、莱斯罗普画廊以及达特茅斯绿地。

College students like to explore the new university world they find themselves in. They yearn to achieve a sense of belonging by becoming the *inside-dopesters*[①] of their environments, the initiates who know where all the secret places are. The new *Hood Museum of Art at Dartmouth*[②](Fig. 26-1) seems made for the purpose of catering to this collegiate and useful impulse. It's a building you could spend your whole four years trying to figure it out and not succeed. You can discern clearly enough the Hood's center—it is the top-floor Churchill P. Lathrop Gallery, dominated by a huge bright Frank Stella collage—but you can't tell at all where its perimeter is. The Hood's edges interdigitate, as the scientists like to say, with everything around it and especially with two older buildings on either side. No matter how many times you wander through the Hood and its neighbors, in and out and up and down among its numberless entrances and staircases, you never succeed in forming a clear mental image of its essentially indeterminate building. Instead, you begin to think of yourself *as the inhabitant of a graphic by Escher*,[③] doomed to climb forever the staircases of a universe of shifting perspectives.

Fig. 26-1   Hood Museum of Art

The Hood's architects are Charles W. Moore, FAIA and Chad Floyd, AIA, of *Centerbrook Architects*[④](Fig. 26-2) with Glenn Arbonies, AIA, as managing partner. *The Hood is sober for Moore & Co. , with a sense of solidly built fabric and institutional permanence that is rare in their work, yet it retains all their usual invention.*[⑤] A simple but inadequate description of the Hood might go like this. It is a 2-storey (plus basement) building made of red and gray brick, dark *bush-hammered concrete*[⑥], and copper. It plugs a gap between two older buildings at a corner of *the Dartmouth Green*[⑦], the grassy common that is the heart of both the village of Hanover, N. H. and the college. It contains 10 gallery rooms, one auditorium and various support spaces on 5574. 182m^2. It also contains very good collections in both art and archeology.

Unfortunately, that description fails to convey anything significant about the Hood. It fails to note, for instance, that the Hood wandersits site as aimlessly as a lost cow, ignoring the orthogonal grid with which other buildings in Hanover are aligned. The Hood has many entrances, but no main entrance. It shapes three courtyards so irregular that they cannot be experienced as outdoor rooms but seem, rather, to be clearings in an architectural forest. All the Hood's parts look different from one another, they tend to be incomplete in themselves, and

Fig. 26-2 Centerbrook Architects

they collide at random angles and levels. If there is a single governing metaphor, it is that of the Victorian New England mill village. But the Hood is a mill village romanticized into something more like an Italian hill town, picturesquely punctuated at the skyline by theatrical gables. Often charming, this exterior occasionally lapses into confusion. The problem is most serious at the most important place, which is the facade facing Dartmouth Green.

Since the Hood itself is set well back from the Green, Floyd and Moore have tried to give it a presence there by means of a false front. This is a ceremonial gateway made of darkened, bush-hammered concrete, lacking detail or ornament and facing north. The architects have wanted granite here, real blocks of solid granite from the Granite State. But the budget will have allowed only thin granite veneer; the architects choose concrete instead. From any distance out on the green, you hardly notice the gate; the Hood instead presents itself as a shapeless heap of copper roofs in the middle distance between and behind its two neighbor buildings. When the roofs turn green the museum will read as a Bosque. As you approach the Hood from the green and notice the gate, you also perceive the museum's name boldly carved into it. This named gateway is an architectural promise that if you pass through, you will arrive at the Hood. Also, you find yourself only in a mysterious courtyard, facing two ramps, one curving up, and one going straight down. I watch as one unfortunate elderly couple starts down, then backtracks and starts up, then gives up and departs.

Having persevered, they will have found that both ramps lead to entrances at different levels, but they have no way of knowing that. Moore and Floyd here play a little too aggressively the game of setting up expectations only to undercut them. Failing to provide a legible entrance is coy. And the split-level courtyard that results, intended for future sculpture, is an awkward and unpleasant space.

Other parts of the Hood's exterior are more prepossessing. The copper roofs, to be sure, are a little unsettled, sloping and interesting in too many ways, like a flight of copper-colored paper airplane. But the Hood nevertheless achieves a kind of wholeness through the use of a consistent palette of materials. The red brick walls with punched, mullioned windows are enjoyable like-but-unlike traditional New England vernacular. *The gray brick, stack-bonded in friezes, helps belt together the diverse shapes.* [8] And the three colors of green-plain green, blue-green and olive green—brighten the eaves and windows, echoing the future color of the copper.

The Hood's interior, not its exterior, is where the big architectural successes are scored. Almost everything inside works well, and the whole sprawling complex comes together with a satisfying crescendo at the Lathrop Gallery, one of the remarkable recent rooms in American architecture. It is this room—just as Floyd and Moore predicted before the building was finished—that is the memorable experience at the Hood. The approach is dramatic: you climb a stair that hugs a glass wall, a wall double-glazed in such a way that each layer has its own separate grid of mullions, so that the grids slide past one another as you move. Light fixtures in the shape of flaming torches line this stair, as if a medieval banquet awaited you at the top. What does await you is a gallery crossed at its gabled ceiling by a catwalk, above which a skylight spills indirect daylight down onto the gallery walls. The room is dominated by the 13ft × 17ft Stella metallic collage, on loan from the artist, a work which radiates energy from the far wall with verve equal to that of the architecture.

Standing in the center of the Lathrop Gallery, you realize you are at the focal point of two separate axial vistas, each of which penetrates a series of galleries. The vistas extend outward from the Lathrop at angles, like two trains departing from a station in different directions. You feel yourself to be at the center of something, in a space that commands and magnetizes everything around it. The Hood interior offers many further delights. All the rooms and passageways are treated differently. As Moore has noted in a published statement: "Among my favorite museums are the little ones with lots of special places—like the Phillips Collection in Washington; I wanted this museum to be a series of rooms of very different proportions, grandeurs and characters, where the art would not just appear in some anonymous matrix but have the opportunity to enjoy its own environment."

The Hood interior succeeds in the way Moore intended, as a linked grouping of individual places. If a building is a family of rooms, as Louis Kahn suggested, then the Hood is a sort of reunion of the members of a very idiosyncratic family, some of whom have chosen to arrive for the festivities in costume. The auditorium, for instance, is a game of hide and seek played with columns. A row of three real, freestanding columns runs down one side. *The middle column lacks its base and most of its shaft, which would have impinged on the seats: Its capital hangs like a severed head, suspended from the ceiling.*[9] On the opposite side, three corresponding column shapes are painted in silhouette on the wall, as if they were shadows of the

first three but with the middle one now complete. All six "columns" serve as light fixtures.

The entry vestibule is another pleasing anomaly. Here several circulation paths come to resolution at a little octagonal circus with a ceiling of blue beams, an arbor-like place that opens out into the nearest thing the Hood has to a main lobby. Everywhere you go in the Hood, details like stair rails and lights are freshly invented. And the installations of the art and artifacts, too, are exceptionally well done, kept carefully in scale with the architecture. Rooms seem made for their contents, contents for their rooms. To say confusion, finally to discuss the two neighboring buildings. On the Hood's west is the Hopkins Center, a building of the 1950s by Harrison & Abromovitz, which contains a theater, a cafeteria. The Hopkins shares with the Hood one quality, which is explorability.

One of the best things Moore and Floyd have done is reaching deeply into the Hopkins Center and renovating some of it into a new cafeteria, relocated next to the entrance to the Hood. An unusual design process produced the Hood, greatly influencing its final form. Centerbrook began work by setting up an office in the Hopkins Center, next to the college snack bar. Even so basic a decision as where to site the museum came out of this collaborative client-architect process. Centerbrook simply designed six different sites, and the committee picked a winner.

The Hood is a building of wonderful parts that seems to wish to convey the sense of an order won momentarily. Like a college student's room, the museum is filled with shrines and sacred places while remaining, as a whole, not fully formed. In its indeterminacy it perhaps resembles the personality of its senior architect, Charles W. Moore, who in the days prior to the Hood's opening was, as usual, dodging around the world on airplanes (Brazil, Berlin), and while simultaneously building in Texas the eighth of the houses he has designed for himself. The Hood possesses Moore's own trait of a nervously amused inventiveness that never quite risks coming to closure. *It is a measure of the high quality of this building that one intensely wishes for it the very greatness it seems perversely resolved to fall just short of.* [10] The Hood Museum is a building which, once seen, lodges forever in the memory, growing in interest the longer you think about it.

Vocabulary and Expressions：

Churchill P. Lathrop Gallery ［ˈtʃəːtʃil piː ˈlæθrɔp］丘吉尔·P·莱斯罗普画廊

Frank Stella ［frænk ˈstelə］n. 弗兰克·斯特拉（美国极简抽象主义画家）

collage ［kəˈlaːʒ］拼贴画

Charles W. Moore ［tʃaːlz ˈdʌbljuːˈmɔːə(r)］n. 查尔斯·W·摩尔（建筑师）

Chad Floyd ［tʃæd flɔid］n. 查德·弗洛伊德（建筑师）

Glenn Arbonies ［glenˈaːbəniəs］n. 格伦·阿伯尼斯（建筑师）

the Granite State 花岗岩之州（新罕布什尔州别名）

Vernacular ［vəˈnækjulə］n. 民居，乡村民宅，民间风格

stack bond（stack-bond）横竖通缝砌法，对缝砌法

the Phillips Collection in Washington 华盛顿菲利普斯艺术博物馆（收藏馆）

Louis Kahn ［ˈlu(ː)is kaːn］n. 路易斯·卡恩（建筑师）

Gothic capital 哥德式柱头

the Hopkins Center 霍普金斯中心

Harrison ［ˈhærisn］n. 哈里森（建筑师）

Abromovitz ［ˌæbrəˈməuviz］n. 阿布拉莫维茨（建筑师）

Escher ［ˈeʃə］n. 埃舍尔（荷兰版画大师）

Notes：

① inside-dopester 熟悉内情者。

② The Hood Museum of Art at Dartmouth College, owned and operated by Dartmouth College, is located in Hanover, New Hampshire (or NH, or N. H.)（汉诺威镇，新罕布什尔州），USA and was founded in 1972. It is one of the oldest and largest college museums in the country. The award-winning building designed

by Charles W. Moore（查尔斯·W·摩尔）and Chad Floyd（查德·弗洛伊德）of Centerbrook Architects（森特布鲁克或中心溪流建筑师事务所，see next page）was completed in 1985, yet the museum's collections stretch back to 1772, three years after Dartmouth College was founded. The collections of the Hood are rich, diverse and available for the use of both the college and the broader community. Access to works of art is provided through permanent collection displays, the online collections database, special exhibitions, the Web site, scholarly publications and programs and events.

③ as the inhabitant of a graphic by Escher 像是囿于埃舍尔版画艺术之中。

④ Centerbrook Architects 森特布鲁克建筑师事务所。Centerbrook was conceived in 1975 as a community of architects working together to advance American place-making and the craft of building. From the beginning, its work has spanned from planning and architecture to details that make buildings memorable. The firm's 19th-century compound of mill buildings, a former wood boring bit factory, is both the firm's home and its vital center of experimentation where design is enriched by many streams of influence. An in-house theater hosts lectures and symposia on many topics, including how to make places distinct and how to perfect the craft of building. A collaborative firm with an exceptional history of building, Centerbrook performs many other services such as project management, planning, interior design, sculpture, landscape and site design, industrial design, furniture and lighting design, fund raising and graphic design. All are done from Centerbrook's historic compound on the Falls River in Centerbrook, Connecticut.

⑤ 摩尔公司精心设计的杰作——护德艺术博物馆，结构坚固，设施耐用，不乏创新。

⑥ bush-hammered concrete 锤凿混凝土。

⑦（the）Dartmouth Green 达特茅斯绿地。Dartmouth Big Green（昵称 Big Green）达特茅斯学院大绿队（昵称大绿队，是一支美国有名的体育运动队）。

⑧ 采用通缝砌筑的青砖饰带将各种各样的

造型结合在一起。

⑨ 中间那根柱子的底座和大部分柱身被悬挂装饰遮挡，甚至还会影响座椅使用：柱头悬挂装饰像一巨幅标题从天花板飞流直下。

⑩ It is a measure of the high quality of this building that one intensely wishes for it the very greatness it seems perversely resolved to fall just short of（省略 which）. 人们强烈希望给这栋高质量建筑的评价是非常宏伟，但好像这个问题解决得不合人意，最后落得的结果是这栋建筑恰恰缺乏这种宏伟。这是一个 It is … that 强调语句，主句宾语为 a measure of the high quality of this building，宾语补足语为 the very greatness，后置定语 it seems perversely resolved to fall just short of（省略 which）修饰 greatness。

Exercises：

1．Translate the Following Paragraphs into Chinese.

The Language of Architecture：Language, as we know it, consists of symbols-words, sentences, gestures, forms-anything that may be put together to communicate. In architecture the symbols are walls, roofs, doors, windows, steps, spires and so on—the element of which buildings are made. Each can be designed in an infinite number of ways and then put together in innumerable variations. Thus the final meaning is an indeterminate intangible which depends, just as does poetry or music, on the creative and expressive power of the creator and the interpretive or receptive capacity of those who respond. Since both ends of this act of communication can be so extremely different, according to the individuals involved-their experiences, prejudices, convictions and sensitivities, it is small wonder that the same piece of architecture can mean so many different things to different people, and be the occasion for so much fierce debate.

2．Translate the Following Paragraph into English.

博物馆作为城市文明符号与主要象征，是重要的公共空间，具有重要的文化价值，能够从物理与心理上拓展城市生活的限制与疆界。博物馆作为展示地区历史文化等内容最为重要的建筑，其公共空间往往是体现地域文化与地域精神的重要空间之一。这种空间由于往往具有复合特点，它是集合着人们休憩、餐饮、娱乐、交流等行为的场所。

27   Team SCUT-POLITO' s LONG-PLAN and Professor' s Green Building Ideas

（华南理工大学－都灵理工大学联队设计的长屋与教授的绿色建筑理念）

教学目标：本单元的主要内容有华南理工大学-都灵理工大学联队、美国能源部太阳能十项全能竞赛、长屋设计方案及特点、生态核心（中庭雨水绿墙、鱼菜健康绿墙）、节能、模块化施工以及绿色建筑理念。教学重点是绿色建筑理念，难点是生态核心（中庭雨水绿墙、鱼菜健康绿墙）。通过本单元的学习，要求熟悉并理解生态核心设计理念，掌握长屋设计方案及特点，了解模块化施工。

27.1   Introduction

Team SCUT-POLITO is composed of teachers and students from South China University of Technology and the Italian Polytechnic University. The team members include top instructors in urban design, green building, energy-saving technology, and undergraduate and graduate students from more than a dozen departments from the two universities, including the School of Architecture, School of Civil and Transportation, School of Electric Power and School of Energy and Environment. As the main body of the team, under the guidance of the teachers, students will give full play to the advantages of creative and multidisciplinary cooperation between the two countries to jointly build a livable, economical zero-energy, ecological, long-narrow house. The LONG-PLAN is the wisdom of both parties. LONG-PLAN, designed by the Team SCUT-POLITO, is a low-rise, high-density, low-energy, long-narrow house that is tailor-made for young and emerging families in high-speed cities. LONG-PLAN has carefully considered the architectural design issues through system integration, and has also responded well to the urban housing problem represented by Guangzhou, bringing us a more ecological, humanistic and technological life experience.

27.2   Competition Plan

The LONG-PLAN Project enhances the freedom of use of space through the control of architectural order and advocates the idea of regaining natural life. As a long-lasting bond that maintains the feelings among the residents, the LONG-PLAN Project integrates human comfort, energy saving, urban effects and module construction. These four strategies to return the life of ecology, humanities and technology to the city and let history go to the future.

27.2.1   Human Comfort

Passive Space Design: The sunlight room and the "Ecology Core" of multiple patios embedded in the long-narrow houses introduce natural air, moisture, sunlight, soil and various creatures, turning the originally closed dark space into a comfortable and livable place, bringing vigor to the drying urban living.

Ecological Core: In the summer, the "Ecology Core" (the patio skylight) can be open, and the HVAC system is not needed. The natural ventilation is comfortable and pleasant. In the winter, the "Ecology Core" (the patio skylight) is closed, and with the high-performance enclosure insulation structure, by collecting the heat of the sun, the energy consumed for heating could be greatly reduced. Passive technology means reducing energy consumption, while active technology means initiatively making use of external resources (Fig. 27-1). The best bet is combining passive technology with active technology.

We are using landscape technologies to combine ecosystem with the house. Semi-exterior vertical garden that uses rainwater, aquaponic system that serve family health, and community-level rainwater gardens are essential elements of urban residential ecology, the whole house system forms a complete rainwater cycle that accelerates the rate of ecological carbon uptake (Fig. 27-2).

A Variety of Spatial Combination Modes: The design of the service-integrated belt provides greater freedom for space use and internal layout. According to the needs of different time periods and people, the long house can be conveniently arranged into a variety of spatial combination modes.

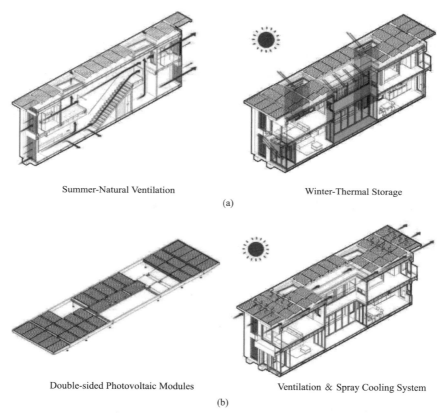

Summer-Natural Ventilation   Winter-Thermal Storage

(a)

Double-sided Photovoltaic Modules   Ventilation & Spray Cooling System

(b)

Fig. 27-1   Passive Technology and Active Technology of the LONG-PLAN
(a) Passive Technology of the LONG-PLAN; (b) Active Technology of the LONG-PLAN

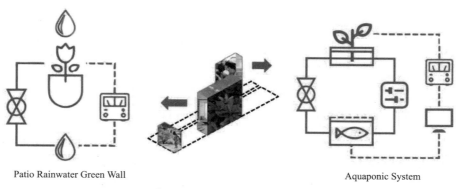

Patio Rainwater Green Wall   Aquaponic System

Fig. 27-2   *Patio Rainwater Green Wall*[1] and *Aquaponic System*[2]

27. 2. 2   Energy Saving

High Efficiency Energy Source: LONG-PLAN's efficient energy system is based on the photovoltaic system and integrated architecture design. Double-sided PV modules and sprinklers are also used in the long house to improve energy system efficiency.

Comfort HVAC System: LONG-PLAN uses small integrated high efficiency HVAC system to avoid the many performance issues associated with fragmented, limited equipment space unique to long-narrow house typology. This system not only ensures pleasant interior environment but also reduces the energy consumption of the house.

Smart Housekeeping: LONG-PLAN intelligent housekeeper masters a machine learning algorithm, which calculates the most energy-efficient equipment operation mode and automatically regulates it through real-time data and user feedback. Perfect and convenient intelligent interactive system, suitable for young people's life scenarios, intelligent energy management system allows residents to conveniently control the house' shydropower equipment (Fig. 27-3).

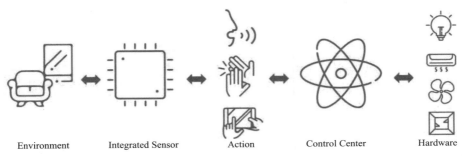

| Environment | Integrated Sensor | Action | Control Center | Hardware |

Fig. 27-3   LONG-PLAN Intelligent Housekeeper

27. 2. 3   Urban Effects

Water Circulation System: The LONG-PLAN's water system takes into account the close integration with urban water networks. The domestic sewage is classified and reused on the community scale, and the rainwater garden in front street is effectively slow down and naturally purifies the rainwater. Efficient small-scale purification equipment is designed and developed by the team to recycle the rainwater, living grey water and air-conditioning condensate from the roof to the hydration, flushing and photovoltaic cleaning of the housing ecosystem.

Community Building: LONG-PLAN has given full play to the advantages of low-rise residences, focusing on the most dynamic neighborhood atmosphere while creating a comfortable family life. The neighborhood interaction in the front street backyard, the public space in the community park, and the energy management of electric vehicle have all contributed to the LONG-PLAN community.

Multi-Mode Development: According to its own structural characteristics, the LONG-PLAN creatively proposes two modes of development: single placement and tract development, and uses elastic transformation to deal with the complexity of urban renewal, the volume ratio of tract development can be as high as 1. 5 ~ 2. 0.

27. 2. 4   Module Construction

Unit Adaptation: The rational modular integrated design ensures efficient operation and free expansion of the whole house to meet different needs of users. The division of the module is perpendicular to the length of the house, making the expansion of the house both horizontally and vertically easier. The size of the module is 3m×4.8m, facilitating factory pre-fabrication and transportation, and has good adaptability.

Integrated Module Wall: The integrated servant wall is confined to one side of the house. More of the service space is vacated, and the space in the depth direction is utilized more efficiently. The main service facilities are placed in the integrated wall, and the comfort of the service space is ensured (Fig. 27-4).

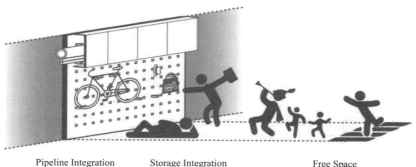

Pipeline Integration   Storage Integration   Free Space

Fig. 27-4   Integrated Module Wall

Modular Furniture: The choice and design of the furniture complies with the basic modulus of 150mm, making the interior furniture economical and flexible. Reasonable modular system allows furniture to be created in a variety of combinations to meet different needs of space use (Fig. 27-5).

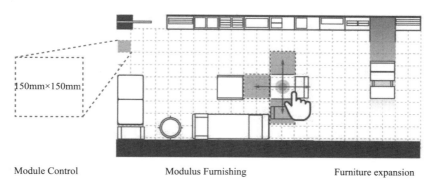

150mm×150mm

Module Control   Modulus Furnishing   Furniture expansion

Fig. 27-5   Modular Furniture

27.3   Professor's Green Building Ideas

As an advisor of 2020 Solar Decathlon Middle East Competition, prof. Yiqiang Xiao (Faculty of Architecture, South China University of Technology) began to pay attention to the influence of climate factors on architecture when he was studying for his master's degree. In the late 1990s, he went to Germany for further study. Germany at that time had generally begun to design green building, so he predicted that green building would become a big trend sooner or later. In the 1980s, it was very popular to use double facades

(Fig. 27-6) in architecture design. In most students' graduation design, you can see a double facade on the building, because in their eyes, the double facade is the most advanced thing. Therefore, some professors say to their students, "Double facade is ok, but it needs the right reasons. Do not imitate it as fashion." Besides, green building is not just simply a passive building; it can be very active to collect energy rather than to take.

However, our city, which is originally a parasite, relies on large distances for energy, water and supplies. He thinks green building could transform it. In Germany, there is a green group—the Green Party. They constantly promote some green ideas to guide the change of laws, the change of people's lifestyle and the change of the industrial structure of society.

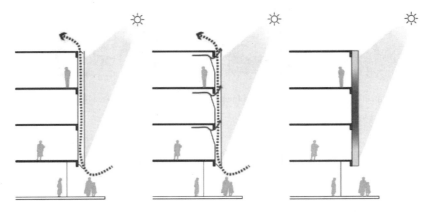

Fig. 27-6   Principles of Double Facade

27.4   Module Lifting and Effect

Module lifting and LONG-PLAN outdoor effect are showed in Figs. 27-7 and 27-8 respectively.

Fig. 27-7   Module Lifting

Green building is a very big concept, not only at the level of architecture, but also the green industry chain, the recycling of materials.

China just began to promote the concept of green building in 2000s. It will take some time to translate it into the social governance, industrial operation and industrial standards of the whole country. However, it is making progress. The study of green building is oriented towards the future, which is to meet the challenges that architecture must meet. He believes that when designing green buildings, design is the premise, not the collage of technology. It is necessary to truly understand what a green building is.

Technology and Art: Prof. Yiqiang Xiao thinks a reasonable view of form, or the so-called artistic pursuit, is the sum of the above. We cannot ignore the history, the venue, or the personality and social feelings of people. When we are teaching, we do not ask everyone to focus on a

Fig. 27-8   LONG-PLAN Outdoor Effect

certain aspect, but to present all the problems. Only students, who are aware of comprehensive problems, can go further in the career of architects. He is in favor of implementing the concept of sustainable energy conservation and environmental protection as a form of architecture, rather than a simple manipulative operation. Form and structure do not exist in isolation. It is necessary to solve the problem of ventilation rationally and ensure the realization of form meanwhile. In fact, the knowledge of engineering is systematic, and is the most basic and easy professional knowledge to master, based on which it is easier to establish confidence in the study of architecture.

Design and Technology: Design is a process of continuous judgment. Generally, we make qualitative judgment with the help of our own or others' experience. For some cases, such as energy consumption, which need more accurate quantitative analysis, the qualitative judgment of experience is not enough. Digital simulation is just a tool that allows architects to predict what the house will be when they devise

a scheme. As an architect, judgment and tool handling are both important. All of our simulation techniques are, in fact, aiding judgment. After the rules are set and the results obtained, it is still up to people to make judgments, because people's judgments have values, emotions and positions. The competition we participate is a simple version of a residential building (Fig. 27-9). In reality, residential buildings face much more complex problems, but the logic is the same. In such a small unit, all the energy resources can work well, which means that the model can be generalized. Simulation is actually a kind of tool, which can help people to extend their brain and can calculate and tell the result, but the judgment is still done by people. Even with simulation, there is no single standard answer. Any choice makes sense.

Reflection and Prospect: As instructors, we do not interfere too much in the competition, but can only point out the direction for students. Therefore, the competition is a test in how to control an entire project with experience while

Fig. 27-9   The LONG-PLAN Model

maintaining the design concept and further development. This is where design skills are tested the most. Every building has its imperfections, and every student needs to experience such regrets. At the end of the competition, we will allow some harmless problems to persist. Although students do have the ability to solve these problems, it takes up a lot of their time. These technical problems will, in turn, test the implementation of creativity. Now we will upgrade our basic strategies on the basis of the former competitions.

The 7 themes of *2020 Solar Decathlon Middle East Competition*[3] are: sustainability, future, innovation, clean energy, mobility, smart solutions and happiness. The actual core of the future green building is to manage space and resources more accurately. It must be a media control technology that relies on the digital age. The intelligent approaches to climate change are also a very important challenge for the future of architecture. What is intelligence? That is, the state of the room can be judged through the computer, and then the user is reminded to take corresponding measures, and even related to the astronomical weather. The system judges itself and implements it directly. This convenience has clearer intelligence in dealing with climate and environmental changes, which is a more accurate reflection of green and energy saving.

The Solar Decathlon provides a good opportunity for students to participate in real construction. Our education actually points to real problem orientation, and a real case is helpful for such education. Provide a site for students to find its problems and try to solve it is the so-called real-problem-orientation education model. The biggest difficulty in architecture education is that we have practiced a lot of skills and thoughts, but it is difficult for us to figure it out 1 : 1. Many architects study architecture in constructing buildings instead of drawing in college. During the whole process, they communicated with craftsmen and supervised the work. So, participating in the competition is a good opportunity to figure out a building in 1 : 1

scale. However, we still need to draw a model to simulate it. It takes a lot of money and time to build a house, so great deviations are not allowed. All the tools are helping to do this. We should make the simulation design as close as possible to the actual project implementation, so the engineering standpoint is very important, because the house is built to use, not drawn on the paper for aesthetic appearance.

Vocabulary and Expressions：

Team SCUT-POLITO 华南理工大学和意大利都灵理工大学联队

The U. S. Department of Energy Solar Decathlon (SD) 美国能源部太阳能十项全能竞赛

HVAC (Heating, Ventilating and Air Conditioning) system 暖通空调系统

photovoltaic system 光伏系统

parasite ['pærəsait] n. 寄生体

digital simulation 数字模拟

Notes：

① Patio Rainwater Green Wall: Patio Rainwater Green Wall is a planting vertical garden using fiber cloth technology and is the core and focus of daily life.

② Aquaponic system: Fish waste makes the water body rich in nutrients. The plants absorb and filter the water and return it to the aquarium to complete a cycle.

③ The U. S. Department of Energy Solar Decathlon (SD) is an award-winning program that challenges collegiate teams to design, build and operate solar-powered houses that are cost-effective, energy-efficient and attractive. Known as the Decathlon competition, it evaluates each home based on the performance in 10 contests. SD aims to promote industry-university-research cooperation in the industry and to facilitate innovation and intensive adoption of solar energy technologies. Since the launch of SD in 2002, subsequent competitions have been hosted in U. S. , Europe, China and Middle East. Over 100 collegiate teams around the world have participated in these competitions. The competition displays scientific research and applications of the participating countries and regions in solar energy, as well as their new energy technologies and achievements in energy conservation and emission reduction.

Exercises:

1. Translate the Following Paragraph into Chinese.

The U. S. Department of Energy Solar Decathlon(SD) is an award-winning program that challenges collegiate teams to design, build and operate solar-powered houses that are cost-effective, energy-efficient and attractive. Known as the Decathlon competition, it evaluates each home based on the performance in 10 contests. SD aims to promote industry-university-research cooperation in the industry and to facilitate innovation and intensive adoption of solar energy technologies. Since the launch of SD in 2002, subsequent competitions have been hosted in U. S., Europe, China and Middle East. Over 100 collegiate teams around the world have participated in these competitions. The competition displays scientific research and applications of the participating countries and regions in solar energy, as well as their new energy technologies and achievements in energy conservation and emission reduction.

2. Translate the Following Paragraph into English.

长屋计划的水系统充分考虑了与城市水网的紧密结合。在社区尺度上将生活污水分类回收利用，前街的雨水花园有效滞缓与自然净化雨水。高效的小型净化设备由团队自行设计研发，将屋面的雨水、生活污水、空调冷凝水回收利用于房屋生态系统的补水、冲厕及光伏清洁降温。

28　3D Printing
（3D 打印）

教学目标：本单元的主要内容有三维打印简介、三维打印书屋和混凝土农宅的创作设计原型、所使用的建筑材料、表面肌理、三维打印施工过程以及建筑内饰和外观效果。教学重点是创作设计原型和表面肌理，难点是三维打印施工过程。通过本单元的学习，要求熟悉并理解三维打印所使用的建筑材料和施工过程，掌握三维打印书屋和混凝土农宅的创作设计原型和表面肌理的设计思想，了解三维打印的发展历史。

28.1 Introduction

We often hear of 3D printing from newscasters and journalists, astonished at what they have witnessed. *A machine reminiscent of the star trek replicator, something magical that can create objects out of thin air.*[①] It can "print" in plastic, metal, nylon and over a hundred other materials. It can be used for making nonsensical little models like the over-printed Yoda, yet it can also print manufacturing prototypes, end user products, quasi-legal guns, aircraft engine parts and even human organs using a person's own cells. *We live in an age that is witness to what many are calling the Third Industrial Revolution. 3D printing, more professionally called additive manufacturing, moves us away from the Henry Ford era mass production line, and will bring us to a new reality of customizable, one-off production.*[②]

Need a part for your washing machine? As it is now, you would order from your repairman who gets it from a distributer, who gets it shipped from China, where they mass-produced thousands of them at once, probably injection-molded from a very expensive mold. In the future, the beginning of which is already here now, you will simply 3D print the part right in your home, from a CAD file you downloaded. If you don't have the right printer, just print it at your local foundry.

3D printers use a variety of very different types of additive manufacturing technologies, but they all share one core thing in common: they create a three dimensional object by building it layer by successive layer, until the entire object is complete. It's much like printing in two dimensions on a sheet of paper, but with an added third dimension: UP, the z-axis. Each of these printed layers is a thinly-sliced, horizontal cross-section of the eventual object. Imagine a multi-layer cake with the baker laying down each layer one at a time until the entire cake is formed. 3D printing is somewhat similar, but just a bit more precise than 3D baking.

While most people have yet to even hear the term 3D printing, the process has been in use for decades. Manufacturers have long used the printers in the design process to create prototypes for traditional manufacturing. But until the last few years, the equipment has been expensive and slow. *Now, fast 3D printers can be got for tens of thousands of dollars, and end up saving the companies many times that amount in the prototyping process.*[③] For example, Nike uses 3D printers to create multi-colored prototypes of shoes. They used to spend thousands of dollars on a prototype and wait weeks for it. Now, the cost is only in the hundreds of dollars, and changes can be made instantly on the computer and the prototype reprinted on the same day.

Some companies are using 3D printers for short run or custom manufacturing, where the printed objects are not prototypes, but the actual end user product. As the printing speed of 3D printers goes up and the prices of them come down, users expect more availability of personally customized products. Even if you don't design your own 3D model, you can still print some very cool pieces. *There are model repositories such as Thingiverse, 3D Parts Database and 3D Warehouse that have model files you can download for free.*[④] What do all these people print? It is limitless. Some print things like jewelry, some print replacement parts for appliances such as their dishwasher, some invent all sorts of original things, some create art, and some make toys for their kids. With the many types of metal, plastic, glass and other materials available (even gold and silver), just about anything can be printed. This is a disruptive technology of mammoth proportions, with effects on energy use, waste, customization, product availability, art, medicine, construction, the sciences and of course manufacturing. It will change the world as we know it.

28. 2 A 3D Printed Book Cabin

A 3D printed Book Cabin made from robots has opened in *Shanghai's Baoshan Wisdom Bay Science and Technology Park*.[5] It is designed by Professor Xu Weiguo of the School of Architecture of Tsinghua University and builds with printing machines and materials developed by the team. It is part of the "Art Bridge" space in Shanghai, which can be used for book shows, academic discussions and book sharing sessions. Professor Xu Weiguo won *"Art Bridge 2021 Annual Figure"*[6]. The design of the Book Cabin starts from the concept sketch (Fig. 28-1), then uses MAYA software according to the needs of entity modeling, and carries out the space shape and structure rationality to determine the implementation model. Then the digital files are created by using the planning and coding of printing path. Finally the digital files drive the robot 3D printing equipment to concrete the materials layer-by-layer stack printing, thus building the curved shape of the Book Cabin (Fig. 28-2). The Book Lodge has a total area of about 30m^2 and can accommodate 13 people for various activities.

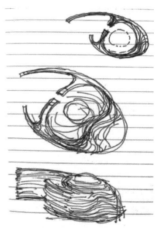

Fig. 28-1　Concept Sketch

Fig. 28-2　Printing Process

The materials used for the Cabin printing is the fiber concrete developed by the team, which does not add steel bars and does not use formwork. The printing of the Cabin uses 2 sets of robot arm printing system, one in situ printing building foundation and main structure, another in situ pre-printing arc wall and dome roof. Each printing equipment needs 2 people to operate, a total of 4 ~ 5 construction technicians to participate in the construction process. The wall structure of the Book Cabin adopts hollow wall design, which is filled with thermal insulation mortar to form thermal insulation wall. The building surface has two kinds of texture, one is laminated surface formed by laminated printing, another is a well-designed woven pattern texture on the side wall in front of entrance, let people have a delicate feeling. The design and construction of the Book Cabin show that 3D printing, as a way of intelligent construction, not only saves materials and manpower, but also has high construction efficiency and high construction speed, and can achieve irregular shape construction and ensure high quality of the construction (Figs. 28-3 and 28-4).

Fig. 28-3   Interior View

Fig. 28-4   Exterior View

28.3   A 3D Printed Concrete Farmhouse

The following Robot 3D printing concrete construction technology is developed by Professor Weiguo Xu Interdisciplinary Team of Tsinghua University. It is based on digital architectural design method and automatic robot control system. The 3D printing technology combined with a special concrete material technology, is a kind of innovative intelligent building technology. This technology has the advantage of man-power saving, high efficiency, low cost, high quality. In addition to realizing the design and construction of traditional farmhouse forms, it can also realize the construction of various beautiful irregular curved shapes. In order to promote the overall revitalization of rural areas and continuously improve the rural living environment, the industrialization of this technology will become a specific and effective measure to improve the quality and modernization level of rural housing.

The Wujiazhuang farmhouse in Hebei Province has a building area of 106m² and uses cave dwelling forms and tube-arch roof structures (Figs. 28-5 and 28-6). Three sets of 3D Printed Concrete Mobile Platforms with Robot Arm are used in the printing construction and the foundation and wall are directly printed in situ. A crane is used to assemble prefabricated tube-arch roofs onto the printed wall. The exterior wall of the building is decorated with woven texture and printed in an integrated way with the structural wall. Thermal insulation materials are poured into the hollow wall to form an integrated exterior wall system of decoration, structure and thermal insulation (Figs. 28-7 and 28-8).

Fig. 28-5 Exterior View of Cave Dwelling

Fig. 28-6 Interior View of Cave Dwelling

Fig. 28-7 3D Printed Concrete Mobile Platform with Robot Arm

Fig. 28-8 Hollow Wall with Woven Texture

The design of architecture, structure, water, heating, electricity, interior decoration and the printing path planning are completed by using the same 3D digital model. The parameterized digital model ensures the consistency of information transmission in the whole design process and the effective coordination of information exchange among various professio-ns. "3D Printed Concrete Mobile Platform with Robot Arm" requires only two people to operate the buttons on the mobile platform to complete the printing and construction of the whole house, which fully integrates and simplifies the concrete 3D printing process. Wujiazhuang Farmhouse in Hebei Province built by 3D printing has reasonable function, beautiful appearance, solid structure and ecological energy saving (Figs. 28-9 and 28-10).

Vocabulary and Expressions:
over-printed Yoda 被套印的尤达(一种玩具)
injection-molded 注塑而成的
foundry ['faundri] n. 代工厂，铸造厂
model repository 模型仓库

Fig. 28-9   Outdoor View of Farmhouse

Fig. 28-10   Indoor View of Farmhouse

disruptive [dis'rʌptiv] adj. 颠覆性的，开拓性的，破坏性的

mammoth ['mæməθ] adj. 巨大的，庞大的

robot ['rəubɔt] n. 机器人

cabin ['kæbin] n. 小屋，客舱，船舱；vt. 把…关在小屋里；vi. 住在小屋里

concept sketch 概念草图

MAYA software MAYA 软件

entity modeling 实体建模

layer-by-layer stack printing 逐层堆叠打印

fiber concrete 纤维混凝土

hollow wall 空心墙体

thermal insulation wall 保温隔热墙体

woven pattern texture 编织图案肌理

intelligent construction 智能建造

Notes：
① 这种机器让人联想到星际迷航中可以凭空创造物体的复制机。
② 我们生活在一个可以目睹许多人称为"第三次工业革命"的时代。3D 打印，或者被更专业地称为附加制造，让我们摆脱了亨利·福特时期的大规模生产线，并进入到新的、可定制的、一次性生产的现实。
③ 现在，快速的 3D 打印机能够以数万美元

的价格获得，并最终在制作原型的过程中，能够为公司节省数倍的花费。
④ 用户可以从诸如 Thingiverse、3D 部件数据库以及 3D 仓库等模型仓库中免费下载它们的模型文件。Thingiverse 是 Makerbot 公司下属的全球最大的 3D 打印网络社区平台。提供免费的 3D 打印设计交流和开源创造社区，为用户提供获取设计方案的渠道，激发消费者对于 3D 打印的兴趣和对家用 3D 打印硬件设备的购买需求。
⑤ 上海宝山智慧湾科技园。
⑥ "艺术之桥"2021 年度人物。

Exercises：

1．Translate the Following Sentence into Chinese and Answer the Following Two Questions in English：

Then through the printing path planning and printing coding to complete the digital file, and then the digital files drive the robot 3D printing equipment to concrete the materials layer-by-layer stack printing, thus building the curved shape of the Book Cabin.

(1) What are part of speech and meaning of the word "concrete" in this sentence?

(2) What is the grammatical function of "layer-by-layer stack printing" in this sentence?

2．Translate the Following Paragraphs into Chinese.

CAD is an acronym. Some people use it for "computer-aided drawing", others for "computer-aided design". There are now many CAD systems available to design organizations. They vary in cost, scope, capability and in suitability for the work of the particular office. All but the simplest systems are intended to be rather more than aids to drawing production. However none is capable of aiding all aspects of the design process. In reality, all systems fall somewhere between "computer-aided drawing" and "computer-aided design". So "CAD" will be used just as a convenient label for a complex technique.

It has taken some time to teach computer to cope with information in graphical form. But the incentives are evident. A design office might typically spend only 5% or 10% of its collective effort on calculations. However the production of drawing might well represent 40% or more of its total workload. So any improvement that can be achieved in drawing efficiency ought to have a high impact. This undoubtedly accounts for much of the huge current interest in CAD.

Therefore the widespread use of computer techniques has had to await the introduction of the interactive graphics systems now labeled as CAD systems. These permit the designer/draughtsman to sit at a workstation and to issue a command and other information to the computer. The computer then performs some function as instructed and displays the result on a graphics screen for the user to check. Assuming all is well, the user can proceed to another step in the process, and another, until the whole design task is completed.

3. Translate the Following Paragraphs into English.

3D 打印机使用了各种各样不同类型的附加制造技术，但是它们都有一个核心的共

同点：它们会通过逐层地建立物体来创建三维物体，直至整个物体完成。这就特别像是在一张纸上进行二维打印，只不过是增加了第三个维度：高度，即 z 轴。

许多人发现术语是他们理解与计算机有关事务的障碍。然而，要完全避免术语的使用是困难的。我们已经遇到了"CAD"这个术语的问题，而且也遇到了另一个广泛使用的词——"系统"。系统这个词特指将多件装置组合在一起形成的计算机构造和使用者用来操作和控制计算机的程序或指令。因此，系统是"硬件"加上"软件"——设备和程序。

使用者与计算机相互配合，与计算机系统之间协同工作。因为计算机擅于从事重复性的工作，具有很大而且特别精确的存储记忆装置，能够迅速完成系统性强的工作且不会疲倦和厌烦。因此，这种协同工作通常是成功的。人类在控制和指导方面具有很大的优越性，可以利用人类的本领，包括判断、经验、直觉、想象和智能来进行这项工作。

29 Fully Grouted Cable Bolts' ANSYS Analysis[①], Construction Techniques and Anchoring Mechanism Theory of Swelling-Rising Concrete Arch with Tree-Root-Shaped Cable Bolt Reinforcement[②])

[注浆锚索的 ANSYS 分析、施工技术及锚固机理（胀锚拱理论）]

教学目标：本单元的主要内容有 ANSYS 计算机数值模拟仿真分析从选择模型、创建模型、划分网格、加载、求解到结果分析的整个研究过程、注浆锚索的施工技术及锚固机理——胀锚拱理论以及英文科技论文的写作技巧。教学重点是注浆锚索的施工技术及锚固机理，难点是 ANSYS 计算机数值模拟仿真分析。通过本单元的学习，要求熟悉并理解注浆锚索的施工技术及锚固机理，胀锚拱理论的主要观点包括膨胀作用、抬升作用和锚索树根筋混凝土拱作用。掌握 ANSYS 计算机数值模拟仿真分析方法，了解英文科技论文的组成与写作技巧。

Although the anchoring technique of fully grouted cable bolts finds a wide application in the reinforcement of slopes, tunnels, deep foundation pit, underground caverns and mine stopes, the rock anchoring mechanism of fully grouted cable bolts in high-temperature environment has not been explored till now. By using advanced engineering simulating analysis software ANSYS, the 3D single-cable and stope models based on the in-situ experiment of reinforcement with fully grouted cable bolts in the high-temperature stope hanging-wall of a copper mine in China are established and analyzed in order to explore the rock anchoring mechanism of a single fully grouted cable bolt and stope models in high-temperature environment. The construction techniques of pre-anchoring fissured stope hanging-wall of some copper mine in China by fully-grouted cable bolts are explored. The construction stages such as drilling anchor holes, making anchorage cables and cement mortar, grouting and curing are expounded. At last, a comprehensive rock anchoring theory of grouted cable bolts in high-temperature environment, i.e., theory of swelling-rising concrete arch with tree-root-shaped cable bolt reinforcement, is presented.

29.1 ANSYS Simulating Analysis of a Single Fully Grouted Cable Bolt

29.1.1 Solving Process

The diameter of a single cable, the outer diameter of grouting cement mortar cylinder, and the outer diameter of the surrounding rock (granite porphyry) cylinder are 0.02m, 0.055m and 1m respectively. The single cable, grouting cylinder and granite porphyry cylinder have the same length 10m. Fig. 29-1 is a 3D single-cable temperature model and Fig. 29-2 shows the meshed elements of the model. Tab. 29-1 lists the mechanical parameters of ANSYS analysis of a single-cable model at 600℃. Assuming that *XOY* surface of Cartesian plane coordinate system is on the outer end surface of the model, the coordinate origin is located in the center of the cable. *Thermal-structural coupled element SOLID98 and direct coupled method are used in order to solve the model*[3]. Gravity inertia load is exerted

in the direction of cable axis. The initial uniform temperature is 25℃. Temperature load 600℃ is exerted on all nodes on the outer end surface of the model. The boundary condition of displacement is that $U_z = 0$ for all nodes on the inner end surface of the model and $U_x = U_y = 0$ for all nodes on the outermost cylinder side surface of the model.

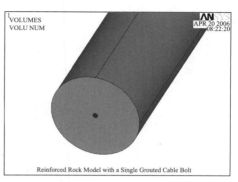

Reinforced Rock Model with a Single Grouted Cable Bolt

Fig. 29-1 A 3D Single-Cable Model

29 Fully Grouted Cable Bolts' ANSYS Analysis, Construction Techniques and Anchoring Mechanism (Theory of Swelling-Rising Concrete Arch with Tree-Root-Shaped Cable Bolt Reinforcement) 〔注浆锚索的 ANSYS 分析、施工技术及锚固机理(胀锚拱理论)〕

213

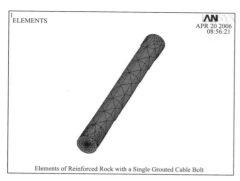

Fig. 29-2 Meshed Elements of a Single-Cable Model

Diametral path *AB* on the cross section through the midpoint in the longitudinal direction of the model, longitudinal paths *CD* through cable center, *EF* through grouting cylinder and *GH* through granite porphyry cylinder are analyzed. The coordinates of the points are $A(0, 0, 5)$, $B(0.5, 0, 5)$, $C(0, 0, 0)$, $D(0, 0, 10)$, $E(0.01875, 0, 0)$, $F(0.01875, 0, 10)$, $G(0.26375, 0, 0)$ and $H(0.26375, 0, 10)$. Solving the model by ANSYS software, we achieve U_z, σ_x and σ_z curves (Figs. 29-3 ~ 29-14) on paths *AB*, *CD*, *EF* and *GH*. Notice that in Figs. 29-3 ~ 29-14 the abscissas are the distance (m) which is from the start point on a path to any point on that path and the ordinates of U_z, σ_x and σ_z curves are $U_z(\text{m})$, $\sigma_x(\text{Pa})$ and $\sigma_z(\text{Pa})$ respectively.

Mechanical Parameters of ANSYS Analysis of a Single-Cable Model at 600℃

Tab. 29-1

Parameters	Granite porphyry	Cement mortar	Cable
Thermal conductivity *KXX* 〔W/(m·℃)〕	0.11	0.19	66.6
Elastic modulus *EX*(Pa)	6×10^9	4.65	1.53×10^{11}
Poisson's ratio *PRXY*	0.3	0.20	0.30
Linear expansion coefficient *ALPX* (/℃)	1×10^{-5}	0.9×10^{-5}	1.06×10^{-5}
Density *DENS* (kg/m³)	2872	2200	7800
Specific heat *C* 〔J/(kg·℃)〕	2.4×10^{-4}	3.9×10^{-4}	460

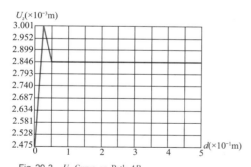

Fig. 29-3 U_z Curve on Path *AB*

29.1.2 Analyzing the Results

From Fig. 29-3, we know that the axial displacement of cable, grouting and rock on diametral path *AB* forms a continuous broken line, and that the maximum displacement difference is only 0.526mm. It is showed that deformation between cable and grouting and that between grouting and rock are harmonious. No staggering movement and no separation occur. Both the cohesive strength between cable and grouting and that between grouting and rock are enough.

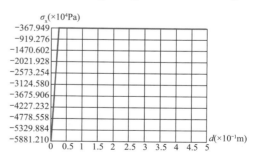

Fig. 29-4 σ_x Curve on Path *AB*

From Fig. 29-4, we see that cable, grouting and rock are compressed in diametral direction. Cable has the maximum compressive stress 58.8121MPa. Grouting and rock almost get the same compressive stress 3.67949MPa. Most stress in surrounding rock is borne by cable. It is what we expect.

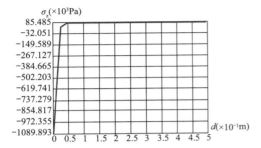

Fig. 29-5　σ_z Curve on Path AB

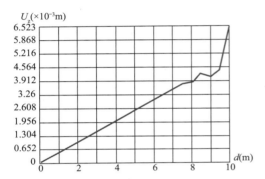

Fig. 29-6　U_z Curve on Path CD

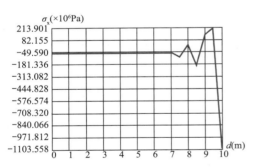

Fig. 29-7　σ_x Curve on Path CD

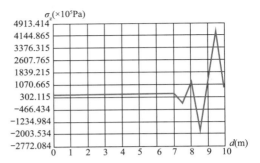

Fig. 29-8　σ_z Curve on Path CD

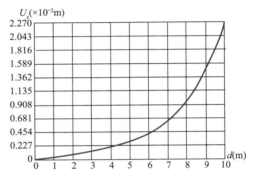

Fig. 29-9　U_z Curve on Path EF

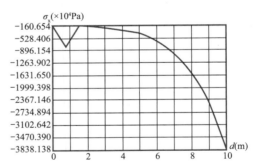

Fig. 29-10　σ_x Curve on Path EF

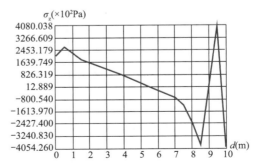

Fig. 29-11　σ_z Curve on Path EF

Fig. 29-5 tells us that on diametral path AB compressive stress in Z direction of cable is only 1.89893MPa and tensile stresses in Z direction of grouting and rock are only 0.026985 ~ 0.085485MPa and 0.085485MPa respectively.

From Figs. 29-6, 29-9 and 29-12, we understand that on axial path displacement U_z of cable increases almost linearly. Both U_z of grouting and that of rock increase along a concave

29 Fully Grouted Cable Bolts' ANSYS Analysis, Construction Techniques and Anchoring Mechanism (Theory of Swelling-Rising Concrete Arch with Tree-Root-Shaped Cable Bolt Reinforcement) 〔注浆锚索的 ANSYS 分析、施工技术及锚固机理(胀锚拱理论)〕

215

parabola. Cable, grouting and rock are all expanded to generate lengthened deformation due to heat-transfer of stope hanging-wall surface. The deformations of grouting and rock are 0.02270m and 0.02238m respectively, and their difference is very small. The deformation of cable is 0.006523m. The deformation difference between cable and grouting is relatively more than that between grouting and rock. Staggering movement and separation between cable and grouting would occur more easily but do not occur actually owing to the fact that all three deformations are very little.

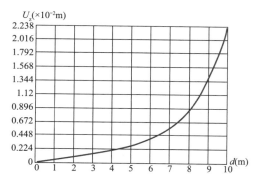

Fig. 29-12 U_z Curve on Path GH

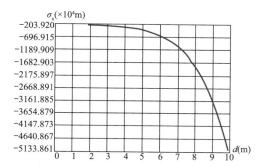

Fig. 29-13 σ_x Curve on Path GH

From Figs. 29-8, 29-11 and 29-13, we know that at the place which is 0.5m away from the mouth of the cable hole in axial direction cable and grouting bear the maximum axial tensile stresses 491.3414MPa and 0.4080038MPa respectively. The maximum axial tensile stress 0.2790337MPa and the maximum axial

compressive stress 0.2201286MPa borne by surrounding rock are located at the bottom of the cable hole and at the place which is 2.5m away from the mouth of the hole in axial direction respectively.

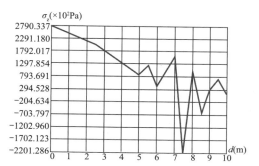

Fig. 29-14 σ_z Curve on Path GH

Because cable grouting and rock cohere firmly in the middle and bottom part of the hole, cable stress is generated mainly by the cohesive force restraining rock deformation. Cable stress in the part close to the stope hanging-wall rock surface is generated mostly by the frictional force restraining rock deformation owing to the fact that the cohesion between cable grouting and rock is not so tight there. From the above principal stress graphs of the single-cable model we can see that a fully grouted cable bolt has an uneven stress distribution in its axial direction. The following two conclusions can be drawn from the analysis. (1) Most stress in surrounding rock is borne by cable. Staggering movement and separation between cable and grouting would occur more easily than those between grouting and rock. (2) The axial displacement of cable, grouting and rock on diametral path forms a continuous broken line, and the maximum displacement difference is only 0.526mm. Deformation between cable and grouting and that between grouting and rock are harmonious. No staggering movement and no separation occur. Cohesive strength between cable and grouting and that between grouting and rock are enough.

29.2   ANSYS Simulating Analysis of a 3D Stope Model

29.2.1   Solving Process

(1) Simulating limits: The horizontal projection length of the stope along inclined direction is 68.79m owing to the pitch angle 30°. Three times horizontal section width along inclined direction is added to the left and right sides in the horizontal direction of the stope individually. So the total simulating width is 143.79m. The embedded depth plus two times vertical height equals the total simulating height 154.75m. Three times strike length of the stope gives the total simulating strike length 21m. Fig. 29-15 is the schematic section of simulating stope limits.

(2) Model selection: 3D simulating ANSYS analysis of high-temperature mine stopes reinforced with fully grouted cable bolts is a thermal-stress coupled problem. Thermal-structural coupled solid element SOLID98 and direct coupled method are used in order to solve the problem. There are three types of materials: granite porphyry, rock reinforced with cable bolts and filling medium. Three normal-temperature stope models are full-fill, half-fill and no-fill. Normal temperature is 25℃. High-temperature stope models have no filling medium. Twelve high-temperature stope models are 35℃, 60℃, 100℃, 200℃, 300℃, 400℃, 500℃, 600℃, 700℃, 800℃, 900℃ and 1000℃.

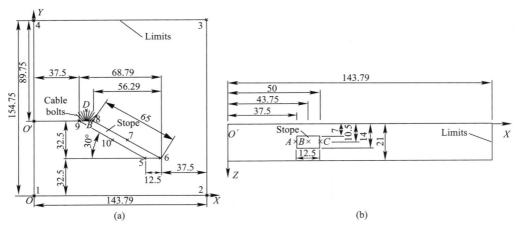

Fig. 29-15   Schematic section of simulating limits (m)
(a) Vertical Section along Inclined Direction through Rear End Surface; (b) Horizontal Section through Stope Hanging-Wall Bottom Surface

(3) Mechanical parameters: Tab. 29-2 shows mechanical parameters of ANSYS analysis of normal-temperature and high-temperature models. Note that for the datum items with parentheses in Tab. 29-2, the data out of and in parentheses belong to normal-temperature and high-temperature models respectively. For the datum items with no parentheses, the data be-

long to all models.

(4) Establishing models: Assuming that XOY surface of Cartesian Coordinate System is on the rear end surface of the model along strike direction and the coordinate origin is located at the left-bottom point of the surface. The positive X direction is horizontal and to the right. The positive Y direction is vertical and up. Ten key

29 Fully Grouted Cable Bolts' ANSYS Analysis, Construction Techniques and Anchoring Mechanism (Theory of Swelling-Rising Concrete Arch with Tree-Root-Shaped Cable Bolt Reinforcement) [注浆锚索的 ANSYS 分析、施工技术及锚固机理(胀锚拱理论)]

217

points are $1(0, 0, 0)$, $2(143.79, 0, 0)$, 3 $(143.79, 154.75, 0)$, $4(0, 154.75, 0)$, 5 $(93.79, 32.5, 0)$, $6(106.29, 32.5, 0)$, 7 $(78.145, 48.75, 0)$, $8(50, 65, 0)$, 9 $(37.5, 65, 0)$ and $10(65.645, 48.75, 0)$ (Fig. 29-15a). Lines, arcs, surfaces and 3D stope models are created by using these key points. Fig. 29-16 is 3D no-fill stope model. Different solid models are meshed with corresponding materials.

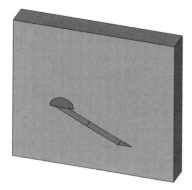

Fig. 29-16   3D no-fill stope model

(5) Loading: Loading of normal-temperature models is steady-state. Length unit is SI (MKS) and temperature unit is Celsius. The initial uniform temperature is $25℃$. Inertia load in Y direction equals acceleration of gravity $9.8m/s^2$. The boundary condition of displacement is that $U_x = U_z = 0$ for all nodes on the two end surfaces of the model along inclined direction, $U_y = U_z = 0$ for all nodes on the bottom end surface of the model along vertical or height direction and $U_z = 0$ for all nodes on the two end side surfaces of the model along strike direction.

Loading of high-temperature models is transient and has 6 loading sub-steps. According to twelve different high-temperature models, the corresponding temperature loads are respectively exerted on all nodes on the two end side surfaces of the mined-out space along inclined direction and on the top and bottom surfaces of the space. Length unit, temperature unit, the initial uniform temperature, inertia load, and the boundary condition of displacement of high-

temperature models are the same as those of normal-temperature models. Fig. 29-17 is no-fill stope model loaded at $1000℃$.

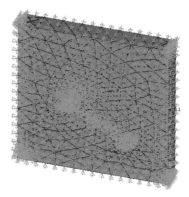

Fig. 29-17   No-fill stope model loaded at 1000℃

(6) Results: Paths ABC and BD are analyzed. Coordinates of points A, B, C and D are $A(37.5, 65, 10.5)$, $B(43.75, 65, 10.5)$, $C(50, 65, 10.5)$ and $D(43.75, 75.5, 10.5)$ (Fig. 29-15). *By solving the models, displacements in Y direction (U_y) and first, second and third primary stresses, i.e., σ_1, σ_2 and σ_3, of the surrounding rock of all models are achieved.* [④] U_y, σ_1, σ_2 and σ_3 curves on paths ABC and BD are also obtained. Fig. 29-18 shows U_y of no-fill model at $800℃$ and Fig. 29-19 gives U_y curve on path ABC of no-fill model at $800℃$. Notice that in Fig. 29-19 the abscissa is the distance $d(m)$ from point A to any point on path ABC and the ordinate is $U_y(m)$. The maximum ground surface displacement U_{ysmax}, U_y, σ_1 and σ_3 at points A, B and C on path ABC, and the maximum and minimum values of U_y on

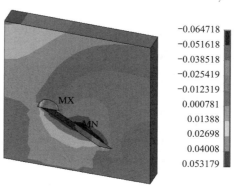

−0.064718	
−0.051618	
−0.038518	
−0.025419	
−0.012319	
0.000781	
0.01388	
0.02698	
0.04008	
0.053179	

Fig. 29-18 U_y of no-fill model at 800℃ (m)

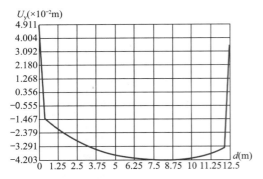

Fig. 29-19 U_y on path ABC of no-fill model at 800℃

path ABC, i.e., max and min, are listed in Tab. 29-3. Notice that in Tab. 29-3 negative and positive displacements stand for subsidence and rising ones respectively, negative and positive stresses means compressive and tensile ones respectively, and units of displacements and stresses are "mm" and "MPa" respectively.

29.2.2   Analyzing the Results

If covering rock thickness under ground surface is between 50 and 500m, the maximum allowable hangingwall subsidence value of hard rock is 20 to 60mm and the allowable ground surface subsidence value, which does not cause public hazards, is 200 to 300mm. Tab. 29-3 shows that maximum subsidence and rising displacements of ground surface are −48mm and 152mm respectively. Maximum hanging-wall subsidence and rising displacements are −45mm and 58mm respectively. Both displacements of ground surface and stope hanging-wall are in their allowable scopes.

The surrounding rock rising displacement caused by heated rock expansion can partly or fully offset the subsidence displacement caused by gravity. Subsidence decrease or even rising displacement of hanging-wall is beneficial to its stability. Hanging-wall rising displacement occurs if stope temperature is equal to or greater than 100℃. Fig. 29-19 and Tab. 29-3 show that rising displacements appear at the corner points A and C of hanging-wall when stope temperature increases. Point B has only subsidence displacement and no rising one. When stope temperature increases the first primary stress of point A changes from the tensile to the

Mechanical parameters   Tab. 29-2

Parameters	Granite porphyry	Reinforced rock	Filling medium
Thermal conductivity [W/(m·℃)]	0.11	0.21	0.16
Elastic modulus(Pa)	$1.53×10^{10}(6×10^9)$	$1.68×10^{10}(1.2×10^{10})$	$2.48×10^9$
Poisson's ratio	0.3	0.25	0.18
Linear expansion coefficient(/℃)	$1.0×10^{-5}$	$1.1×10^{-5}$	$1.16×10^{-5}$
Density(kg/m³)	3340 (2872)	3344 (2876)	2080
Specific heat [J/(kg·℃)]	$2.4×10^{-4}$	$2.1×10^{-4}$	$4.3×10^{-4}$

U_y, σ_1 and σ_3 on ABC and U_{ysmax}   Tab. 29-3

Models	U_{ysmax} (mm)	U_y(mm)					σ_1(MPa)			σ_3(MPa)		
		A	B	C	Max	Min	A	B	C	A	B	C
Full-fill	−37	−11	−25	−29	−11	−29	0.46	0.25	0.05	−1.87	0.03	−0.21
Half-fill	−42	−11	−28	−30	−11	−31	0.57	0.30	0.01	−1.98	0.05	−0.25
No-fill	−48	−11	−29	−31	−11	−32	0.59	0.30	0.01	−2.06	0.06	−0.25

continued

Models	U_{ysmax} (mm)	U_y (mm)					σ_1 (MPa)			σ_3 (MPa)		
		A	B	C	Max	Min	A	B	C	A	B	C
35℃	−42	4	−27	−31	4	−33	2.75	−0.01	0.08	−1.47	−5.12	−1.47
60℃	−17	6	−27	−27	6	−33	2.61	0.01	0.03	−2.39	−8.48	−2.49
100℃	36	8	−27	−25	8	−33	2.42	−0.03	1.25	−4.25	−13.97	−5.10
200℃	55	14	−30	−9	14	−32	2.07	−0.10	1.93	−9.26	−27.33	−10.22
300℃	58	19	−31	3	19	−33	1.89	−0.17	3.13	−14.28	−31.80	−15.25
400℃	69	25	−32	1	25	−35	1.38	−0.43	4.56	−19.31	−38.79	−19.31
500℃	82	31	−34	8	31	−37	1.26	−0.56	5.73	−24.33	−41.40	−24.33
600℃	90	36	−35	44	44	−38	0.97	−0.33	6.68	−29.36	−81.21	−31.43
700℃	105	43	−37	57	57	−40	0.63	−0.39	7.87	−34.39	−94.68	−36.42
800℃	123	49	−38	35	49	−42	0.43	−1.20	14.32	−39.19	−109.22	−39.19
900℃	135	54	−40	56	56	43	−0.33	−0.50	10.24	−44.44	−121.62	−48.2
1000℃	152	58	41	55	58	−45	−0.49	−1.53	18.06	−49.06	−136.44	−49.06

compressive and both compressive stress of point A and tensile stress of point C increase gradually. Point A is mainly compressed, point C has relatively great tensile stress, and subsidence of point B is relatively great. Partial collapse of points B and C is easy to occur.

The maximum tensile stress (7.87MPa) at 700℃ starts to go beyond the allowable tensile stress (6.81MPa) of the anchored rock, and the maximum compressive stress (81.21MPa) at 600℃ begins to exceed the compressive strength (30MPa) of the surrounding rock at the same temperature. Mining is safe if stope temperature is equal to or less than 500℃. For actual temperature of the stope is generally between 35℃ and 300℃, the stope hanging-wall

anchoring parameters can meet the requirements of surrounding rock stability in the process of extraction.

Both displacements of ground surface and stope hanging-wall are in their allowable scopes. The surrounding rock rising displacement caused by heated rock expansion can partly or fully offset the subsidence displacement caused by gravity. Hanging-wall rising displacement occurs if stope temperature is equal to or greater than 100℃. Partial collapse of points B and C is easy to occur. Mining is safe if stope temperature is equal to or less than 500℃. The stope hanging-wall anchoring parameters can meet the requirements of surrounding rock stability in the process of extraction.

29.3   Pre-Anchoring Construction

Anchor holes in the unstable fissured hanging-wall are drilled by YGZ-90 type drilling equipment. [5] The diameter and depth of an anchor hole are 55mm and 10.5m respectively. Upward anchor holes in a row form a vertical fan shape. The corresponding anchor holes in two adjacent rows use staggered arrangement. The row spacing is 3m. Fan-shaped layouts of anchor holes in two adjacent rows are as shown in Fig.29-20. Every anchor hole is cleaned with

high-pressure water or compressed air. An anchor cable uses 20mm diameter steel wire rope with the length of 11m. 6 ~ 8 strands of steel wires are taken out from one end of steel wire rope and are bent into 5 ~ 10cm long a gnail around the outer wire rope as anchor cable head. Finally the steel bar hook (Fig.29-21) is strapped on anchor cable head with iron wire and the anchor cable is pushed into the bottom of the hole by using pushing rods.

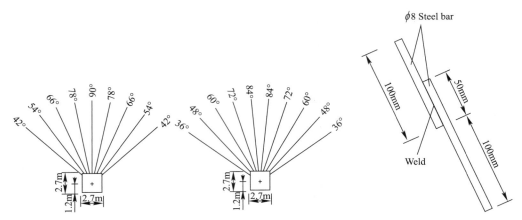

Fig. 29-20 Fan-shaped layouts of anchor holes in two adjacent rows

Fig. 29-21 Steel bar hook

Ordinary Portland cement is used and the largest particle size of sand is not more than 5mm. Cement mortar is made underground and the mixing weight ratio of cement ∶ sand ∶ water = 1 ∶ 1 ∶ 0.3. For each anchor hole 25kg cement and 25kg sand are approximately used. Injection pump is employed for injecting cement mortar into anchor holes. After grouting, anchor hole orifices are plugged with cork, cotton or cement bag paper lest mortar outflow from anchor holes. Mortar strength has

relation to curing time. To assure quality mortar should be cured for at least half a month and the best curing time is a month or more. The deformation of the unstable fissured hanging-wall can be inhibited by the hardened fully-grouted cable bolts combined with fissured hanging-wall. Fig. 29-22 shows the schematic vertical section of the test mine stope and fissured stope hanging-wall pre-anchored by fully-grouted cable bolts along inclined direction.

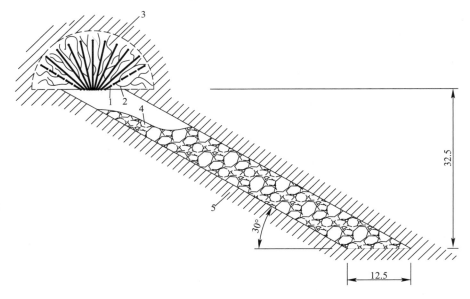

Fig. 29-22 Schematic vertical section of the test mine stope and fissured stope hanging-wall pre-anchored by fully-grouted cable bolts along inclined direction(m)
1-Fully-Grouted Cable Bolts; 2-Faults and Fissures; 3-Fissured Mine Stope Hanging-Wall Pre-Anchored by Fully-Grouted Cable Bolts; 4-Ore; 5-Surrounding Rock

29　Fully Grouted Cable Bolts' ANSYS Analysis, Construction Techniques and Anchoring Mechanism（Theory of Swelling-Rising Concrete Arch with Tree-Root-Shaped Cable Bolt Reinforcement）　［注浆锚索的 ANSYS 分析、施工技术及锚固机理（胀锚理论）］

221

29.4　Theory of Swelling-Rising Concrete Arch with Tree-Root-Shaped Cable Bolt Reinforcement

29.4.1　Model

In terms of anchoring rock with a single fully grouted cable bolt, it looks like the root of a tree called "the tree root of a cable bolt". The cable bolt bound up with cement mortar, grouting in cracks such as faults and soft layers, and grouting in fissures and cleavages are similar to the main trunk, the primary boughs and the secondary twigs of the tree root respectively. The construction process of fully grouted cable bolts is just the growing one of "the tree roots of cable bolts".

The anchoring rock of a single fan-shaped row of fully grouted cable bolts is similar to a reinforced concrete arch bridge. The steel bar in the bridge is called "tree-root-shaped cable bolt reinforcement" (Fig. 29-23). The fractured rock bound up with cement mortar corresponds to concrete, and therefore the anchoring rock is referred to as "concrete arch with tree-root-shaped cable bolt reinforcement" (Fig. 29-24). An arch has two main characteristics: one is that an arch mainly bears compressive force and

the other is that there is great horizontal thrust at the foot of an arch. "Concrete arch with tree-root-shaped cable bolt reinforcement" takes advantage of great compressive strength of rock completely and transmits the loads of stope hanging-wall to surrounding rock. Stress distribution of "concrete arch with tree-root-shaped cable bolt reinforcement" is more reasonable than that of beams.

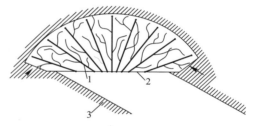

Fig. 29-24　Diagram of Concrete Arch with Thee-Root-shped Cable Bolt Reinforcewent

1-Tree-Root-Shaped Cable Bolt Reinforcement; 2-Concrete Arch with Tree-Root-Shaped Cable Bolt Reinforcement; 3-Surrounding Rock

29.4.2　Main Viewpoints

"Theory of swelling-rising concrete arch with tree-root-shaped cable bolt reinforcement" includes two main viewpoints, i.e., swelling-rising effect and effect of concrete arch with tree-root-shaped cable bolt reinforcement.

29.4.2.1　Swelling-Rising Effect

（1）Swelling Effect

Surrounding rock of underground high-temperature stopes is infinite or half-infinite for embedded depth of the stopes is relatively great. The thermal stress called swelling ground pressure is generated in rock due to the fact that energy of heated rock expansion cannot be released. This is an unfavorable aspect of high-

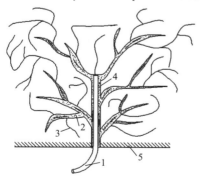

Fig. 29-23　Diagram of "Tree-Root-Shaped Cable Bolt Reinforcement"

1-Cable Bolts; 2-Faults; 3-Fissures; 4-Mortar; 5-Surrounding Rock

temperature swelling effect on anchoring rock.

The closer to the high-temperature stope surrounding rock is, the higher the temperature in the rock is. In high-temperature environment the cable bolt expands and becomes wedge-shaped. The larger end of the wedge is at the cable hole mouth. Both friction force between cable and grouting and cohesive strength of the cable bolt are increased. This is a favorable aspect of high-temperature swelling effect on anchoring rock.

(2) Rising Effect

In high-temperature environment, surrounding rock rising displacement caused by heated rock expansion can partly or fully offset the subsidence one caused by gravity. Subsidence decrease or even rising displacement of hanging-wall is beneficial to its stability. This is high-temperature rising effect on anchoring rock.

29. 4. 2. 2   Effect of Concrete Arch with Tree-Root-Shaped Cable Bolt Reinforcement

Effect of concrete arch with tree-root-shaped cable bolt reinforcement includes the following five main effects.

(1) Hanging effect: Cable bolts are different from bolts. The flexibility of cable bolts is something like that of tree roots. Action of faults can make a cable bolt become curved or appear a broken line, but its anchoring force still exists. In-situ we have occasionally observed that fully grouted cable bolts hang some unstable rock blocks on the stope hanging-wall surface and the rock blocks can not drop. This is the hanging effect of tree-root-shaped cable bolt reinforcement.

(2) *Effect of a laminated beam: Tree-root-shaped cable bolt reinforcement, which is something like linen thread in the sole of cloth shoes, sews up the rock layers of stope hanging-wall together to form a laminated component as a whole. Resisting force between rock layers increases greatly, and the bending strength, shear strength and rigidity of hanging-wall rock also grow.* [6]

(3) Effect of injection anchorage: That cement mortar injects and fills faults, fissures and cleavages makes rock being in a 1D or 2D stress state become in a more stable 3D one. Injection can also increase the strength and deformation modulus of soft constructional surfaces in surrounding rock. Strip and pulse-shaped grouting, which is something like the root hair of a tree, wraps hanging-wall fissured rock into a wedge-shaped thing. This wedge wedges the top part of the stope. The space system of forces of the whole stope is balanced, and therefore no hanging-wall collapse occurs.

(4) *Waterproof and impervious effect: Injection fills faults, fissures and cleavages in surrounding rock to form an artificial impervious hanging-wall with certain thickness. Influence of water on surrounding rock is alleviated or effectively protected. This waterproof and impervious effect is the same as principles of water-protection, sand-fixation and water and soil conservation of tree roots.* [7]

(5) Effect of anchoring in advance: Fully grouted cable bolts reinforce stope hanging-wall in advance before in-situ rock is disturbed. The strength of surrounding rock itself is maintained and acquires further improvement and enhancement owning to the fact that tree-root-shaped cable bolt reinforcement restrains deformation of surrounding rock. It is protected that a 3D stress state of surrounding rock is transformed into a plane stress state in the process of mine extraction. The stability of high-temperature stopes is guaranteed.

Vocabulary and Expressions:
in-situ experiment 现场试验
pre-anchoring [pri:'æŋkəriŋ] n. 预锚固
anchor hole 锚孔
anchorage cable 锚索
cement mortar 水泥砂浆
grouting ['grauting] n. 灌浆，注浆；v. 给…灌浆，给…注浆
ANSYS simulating analysis ANSYS 模拟分析
surrounding rock 围岩
rock anchoring mechanism 岩体锚固机理
granite porphyry 花岗斑岩

mechanical parameter 力学参数

Cartesian plane coordinate system 笛卡尔平面
　　坐标系

cross section 横截面

displacement [dis'pleismənt] n. 位移

longitudinal direction 纵向

abscissa [æb'sisə] n. 横坐标

ordinate ['ɔːdinət] n. 纵坐标

single-cable model 单根锚索模型(单锚模型)

thermal conductivity 导热系数，热导率

elastic modulus 弹性模量

poisson's ratio 泊松比

linear expansion coefficient 线膨胀系数

density ['densiti] 密度

specific heat 比热

deformation [,diːfɔː'meiʃn] n. 变形

tensile stress 拉应力

compressive stress 压应力

concave [kən'keiv] adj. 凹的，凹面的

parabola [pə'ræbələ] n. 抛物线

principal stress 主应力

first, second and third principal stresses 第一、
　　第二和第三主应力

stress distribution 应力分布

parenthesis [pə'renθəsis] n. 圆括号，括弧

strike direction 走向

inclined direction 倾向

subsidence ['sʌbsidns] n. 沉降，沉陷，下沉

maximum allowable hangingwall subsidence val-
　　ue 顶板最大容许沉降量

agnail ['æg'neil] n. 倒刺

fault [fɔːlt] n. 断层

fissure ['fiʃə(r)] n. 裂隙，裂缝

bending strength 抗弯强度

shear strength 剪切强度，抗剪强度

rigidity 刚度

sand-fixation 固砂作用

Notes:

　　本单元专业词汇较多较偏，内容偏难，涉及学科前沿新理论——"胀锚拱理论"，可作为选学内容。ANSYS 是一款计算机数值模拟仿真分析软件，比较适合于高年级本科生及研究生学习使用。读者不仅可学习到 ANSYS 计算机数值模拟仿真分析从选择模型、创建模型、划分网格、加载、求解到结

果分析的整个研究过程，还可习得一些科技论文的写作技巧。

① 注浆锚索 ANSYS 模拟仿真分析。注浆锚索是先在岩体中钻孔，将锚索送入孔中，使用注浆泵将水泥砂浆压入孔中，达到锚固裂隙岩体的效果。ANSYS 模拟仿真分析是根据实际研究对象，使用 ANSYS 计算机模拟仿真软件建立模型，对研究对象的各种参数如应力应变进行分析，找出规律，供实际参考。

② 注浆锚索锚固裂隙岩体机理——"膨胀抬升锚索树根筋混凝土拱理论"，简称"胀锚拱理论"。该理论的主要观点包括膨胀作用、抬升作用和锚索树根筋混凝土拱作用。而锚索树根筋混凝土拱作用主要包括悬吊作用、叠合梁作用、注浆加固作用、截水防渗作用和预加固作用。"胀锚拱理论"克服了目前各种单方面锚固理论的不足，综合形象地解释了高温环境下岩体锚固机理。研究成果为在高温环境下或可能遭受高温影响的边坡、隧道、基坑、硐室、深井开采和有矿石自燃发火危险的矿山开采等土木工程的设计和施工提供科学决策的依据及有益的参考。

③ 该模型选择 SOLID98 热-结构耦合实体单元，采用直接耦合法进行求解。

④ 通过模型求解得到所有模型 Y 方向的位移 U_y 和围岩中的第一、第二和第三主应力 (σ_1, σ_2, σ_3)。

⑤ 使用 YGZ-90 型钻孔设备在不稳定的破碎顶板钻凿锚孔。

⑥ 叠合梁作用：锚索树根筋就像纳鞋底的细麻绳一样将采场顶板各岩层缝合锁紧形成一个整体的叠合构件，岩层结构面的阻力大大增加，提高了顶板岩层的抗弯、抗剪强度和刚度。

⑦ 截水防渗作用：注浆使顶板围岩内的断层、节理、裂隙得以充塞、密实、加固，形成一定厚度的连续的不透水人工顶板，有效防止或减轻水对围岩的弱化，起到截水防渗的作用。这和树根的防水固砂、保持水土的作用原理是一样的。

Exercises:

1. Translate the Following Paragraphs into Chinese.

　　Both displacements of ground surface and

stope hanging-wall are in their allowable scopes. The surrounding rock rising displacement caused by heated rock expansion can partly or fully offset the subsidence displacement caused by gravity. Hanging-wall rising displacement occurs if stope temperature is equal to or greater than 100℃. Partial collapse of points B and C is easy to occur. Mining is safe if stope temperature is equal to or less than 500℃. The stope hanging-wall anchoring parameters can meet the requirements of surrounding rock stability in the process of extraction.

Swelling Effect：Surrounding rock of underground high-temperature stopes is infinite or half-infinite for embedded depth of the stopes is relatively great. The thermal stress called swelling ground pressure is generated in rock due to the fact that energy of heated rock expansion cannot be released. This is an unfavorable aspect of high-temperature swelling effect on anchoring rock.

The closer to the high-temperature stope surrounding rock is, the higher the temperature in the rock is. In high-temperature environment the cable bolt expands and becomes wedge-shaped. The larger end of the wedge is at the cable hole mouth. Both friction force between cable and grouting and cohesive strength of the cable bolt are increased. This is a favorable aspect of high-temperature swelling effect on anchoring rock.

2. Translate the Following Paragraphs into English.

锚索树根筋就像纳鞋底的细麻绳一样将裂隙岩体缝合成一个整体。锚索树根筋混凝土拱的这种叠合作用使顶板各岩层锁紧形成叠合构件，岩层结构面的阻力大大增加，提高了顶板岩层的抗弯、抗剪强度和刚度，从而大大增加了顶板岩层的稳固程度。

水泥砂浆充填节理、裂隙等不连续面的间隙，使间隙两侧处于单向或双向应力状态的岩体进入三向应力状态，更加稳定。注浆还可提高围岩弱面的强度和变形模量。当注浆压力大到一定程度时，浆液流动使岩层产生水力劈裂，形成脉状或条带状胶结体，像树根分枝小须一样，将顶板破碎围岩包裹在一起形成一个大头朝上小头朝下的楔形体，这种注浆加固作用有效控制了顶板的冒落。

水对围岩有软化、溶蚀等作用，能显著降低围岩的强度。围岩注浆封堵了裂隙和流水通道，可有效防止围岩风化，防止或减轻水对围岩的弱化。注浆使裂隙岩体内的断层、节理、裂隙得以充塞、密实、加固，形成一定厚度的连续的不透水人工顶板，起到截水防渗的作用。这和树根的防水固砂、保持水土的作用原理是一样的。

3. Writing Skills of Specialized English

写作是运用语言技巧来表达完整的思想，是英文学习中的一项重要内容，而写作技巧的应用与语言的优美性是英文写作成功与否的决定因素。由于专业英语文体的主要功能是叙述科技事实、记录科技知识和描述科技新发现，其语言特征是结构严谨和客观准确。

科技文章崇尚严谨周密，要求行文简练且重点突出。尽量做到严谨中有变化、句子结构多样化（如采用长短句结合、排比句型、主从复合句型、强调句型等）、表达方式强弱有变化以及适当运用修辞手法。使用强调句型可以避免平淡，加强表达新观点。

专业英语写作中常采用的文体有两种：一种是以主题句为直线展开的。主题句常出现在文章或段落的开头，其他内容往往是对它的补充或解释。行文中可用大量的短句让描述内容逐层递进；另一种是按叙述的内容逐层展开的文体。本文"Fully Grouted Cable Bolts' ANSYS Analysis, Construction Techniques and Anchoring Mechanism（Theory of Swelling-Rising Concrete Arch with Tree-Root-Shaped Cable Bolt Reinforcement）"主要是按科研内容逐层展开的。首先按"a Single Fully Grouted Cable Bolt""a 3D Stope Model""Pre-Anchoring Construction"和"Theory of Swelling-Rising Concrete Arch with Tree-Root-Shaped Cable Bolt Reinforcement"等内容展开，然后"a 3D Stope Model"按"Solving Process"和"Analyzing the Results"展开，最后"Solving Process"又按"Simulating Limits""Model Selection""Mechanical Parameters""Loading"和"Results"这样逐层展开的。

一篇研究性的英语专业论文主要包括标题、作者信息、摘要、关键词、正文（引言

29 Fully Grouted Cable Bolts' ANSYS Analysis, Construction Techniques and Anchoring Mechanism (Theory of Swelling-Rising Concrete Arch with Tree-Root-Shaped Cable Bolt Reinforcement) ［注浆锚索的 ANSYS 分析、施工技术及锚固机理（胀锚拱理论）］

225

和论证）、致谢和参考文献等几个组成部分。不同的杂志对文章格式的要求不尽相同，但大同小异。

（1）标题（Title）：大多数读者是通过原始期刊，或通过二次文献（文摘或索引）读到论文标题的。所以标题中的每个词都应仔细推敲，词与词之间的关系也应认真处理。标题需简短明了，不能流于空泛和过于烦琐，逻辑顺序清晰，能概括整篇文章的中心思想。如本文标题是"Fully Grouted Cable Bolts' ANSYS Analysis, Construction Techniques and Anchoring Mechanism (Theory of Swelling-Rising Concrete Arch with Tree-Root-Shaped Cable Bolt Reinforcement)"［注浆锚索的 ANSYS 分析、施工技术及锚固机理（胀锚拱理论）］，很清楚这篇文章是研究"Fully Grouted Cable Bolts"（注浆锚索）的，主要包括三个方面的内容：ANSYS Analysis（ANSYS 分析）、Construction Techniques（施工技术）及 Anchoring Mechanism（锚固机理），还着重把提出的一个新理论——"Theory of Swelling-Rising Concrete Arch with Tree-Root-Shaped Cable Bolt Reinforcement"（胀锚拱理论）放在锚固机理后面，使得文章特点鲜明、引人注目。

英文专业论文的标题一般是一个名词性短语，通常不写成一个句子，也不出现从句，标题的末尾也不加句点。在行文格式上，标题的第一个词和每个实词的第一个字母都要大写，而小词如冠词、介词和连接词不用大写。在确定论文标题时，最重要的是作者应提供能正确表达文章内容的"关键词"，标题用词应容易理解，便于检索，同时又能突显文章重要内容。

（2）作者信息（About Author）：标题之后必须列出作者信息，如姓名、工作单位、地址、邮编及 Email 等，以便交流。论文署名既表示著作权，也表示文责自负。工作单位如果是大学的话，一般写到学院或系一级即可。需要注意的是英文的地址是从小到大的，如市、省、国家。

（3）摘要（Abstract）：研究性英文专业论文的摘要应简短扼要，中心突出，一般在300 字以内。论文摘要概括研究工作的主要目的、研究成果、采用方法和主要结论，必要时加一两句概述研究背景、重要性和意义。摘要绝不应该提及论文中没有涉及的内容或结论。本文的第一段其实就是整篇文章的论文摘要，只有 173 个英文单词：

Although the anchoring technique of fully grouted cable bolts finds a wide application in the reinforcement of slopes, tunnels, deep foundation pit, underground caverns and mine stopes, the rock anchoring mechanism of fully grouted cable bolts in high-temperature environment has not been explored till now. By using advanced engineering simulating analysis software ANSYS, the 3D single-cable and stope models based on the in-situ experiment of reinforcement with fully grouted cable bolts in the high-temperature stope hanging-wall of a copper mine in China are established and analyzed in order to explore the rock anchoring mechanism of a single fully grouted cable bolt and stope models in high-temperature environment. The construction techniques of pre-anchoring fissured stope hanging-wall of some copper mine in China by fully-grouted cable bolts are explored. The construction stages such as drilling anchor holes, making anchorage cables and cement mortar, grouting and curing are expounded. At last, a comprehensive rock anchoring theory of grouted cable bolts in high-temperature environment, i.e., theory of swelling-rising concrete arch with tree-root-shaped cable bolt reinforcement, is presented.

第一句"Although the anchoring technique of fully grouted cable bolts finds a wide application in the reinforcement of slopes, tunnels, deep foundation pit, underground caverns and mine stopes, the rock anchoring mechanism of fully grouted cable bolts in high-temperature environment has not been explored till now."讲的就是研究背景，虽然注浆锚索应用广泛，但高温环境下的锚固机理尚无人问津。

中间三句"By using advanced engineering simulating analysis software ANSYS, the 3D single-cable and stope models based on the in-situ experiment of reinforcement with fully grouted cable bolts in the high-temperature stope hanging-wall of a copper mine in China are established and analyzed in order to explore the rock anchoring mechanism of a single fully grouted cable bolt and stope models in high-temperature environment. The construction techniques of pre-anchoring fissured stope hanging-wall of some copper mine in China by fully-grouted cable bolts are explored. The construction stages such as drilling anchor holes, making anchorage cables and cement mortar, grouting and curing are expounded. "讲的是研究内容和方法，研究内容主要有建立模型、预加固施工工艺过程和锚固机理，研究方法有 ANSYS 模拟仿真分析、现场试验等。

最后一句"At last, a comprehensive rock anchoring theory of grouted cable bolts in high-temperature environment, i.e., theory of swelling-rising concrete arch with tree-root-shaped cable bolt reinforcement, is presented. "讲的就是研究成果，提出了一个新的理论——胀锚拱理论。

（4）关键词（Key Words）：关键词是指从论文的正文、摘要中抽出的，在表达论文内容、主题等方面具有实际意义并起关键作用的词汇。应尽可能采用规范化的专业词汇，大多是名词性术语，部分是具有检索意义的动词和形容词等。一篇文章一般选用 3 ~ 8 个关键词。每个关键词之间用分号隔开，最后一个关键词之后不加标点。如本文的关键词可取为：grouted cable bolts；ANSYS simulating analysis；anchoring mechanism；theory of swelling-rising concrete arch with tree-root-shaped cable bolt reinforcement（注浆长锚索；ANSYS 仿真分析；锚固机理；胀锚拱理论）。

（5）引言（Introduction）：引言是论文的开场白，是正文的开头部分。它向读者交代本项研究的来龙去脉，对论文的总体轮廓作出概述。要求言简意赅、重点突出、开门见山、紧扣主题、实事求是、不用客套。主要包括研究的目的背景、文献综述、理论依据、试验基础、研究方法、预期成果和作用意义等。

（6）论证（Proof）：论证是一篇论文的主体，在论文中占主要篇幅。可依据逻辑顺序分为若干小部分，每一小部分论证一个或多个结论。总体要求是：论点明确，论据充分，论证合理；事实确凿，数据准确，计算无误；文字简练，条理清楚，层次分明；图表、数量单位、标点、字母大小写等使用规范；正确引用参考文献。

论证部分应该描述试验过程并提供详细的试验细节，以使有能力的研究人员可以重复这个试验。当你的论文受到同行们的审核时，一个好的审稿人员会仔细阅读这个部分。若对你的试验能否被重复做表示严重质疑，无论你的研究成果多么令人鼓舞，都会建议退稿。

（7）致谢（Acknowledgement）：该部分不是必需的，有时根据需要写上一个致谢的语句，感谢对本研究作出指导或重要帮助的师友，言辞恳切、实事求是、简明扼要、恰如其分。

（8）参考文献（References）：它是论文的必要组成部分，反映作者严谨的科学态度和对引用他人成果的尊重。参考文献在正文中应标记有相应的索引，并在全文最后的参考文献表中依序列出。参考文献的格式应符合出版社的要求，准确地写出文献的作者、文献名称、文献所载的期刊名（或书名或论文集名称等）、杂志期卷、出版年、页码（或书的出版地点、出版社、出版年）。举例如下：

参考杂志：P. Naess. Urban Planning and Sustainable Development[J]. European Planning Studies. 2001, 9(4)：503-524.

参考书籍：E. P. Popov. Mechanics of Materials[M]. 2nd ed. London：Prentice/ Hall International, Inc., 1978.

Appendix 1: The Reading of Common Symbols and Mathematical Formulas
（附录 1：常用符号和数学公式的读法）

1. Common Punctuation Marks（常用标点符号）（Tab. 1）

Common Punctuation Marks（常用标点符号）   **Tab. 1**

Mark（符号）	English name（英文名称）	Chinese name（汉语名称）
.	period（or full stop，full point）	句号
,	comma	逗号
;	semicolon	分号
:	colon	冒号
?	question mark（or interrogation point）	问号
!	exclamation mark（or exclamation point）	感叹号（或感情号，惊叹号）
" "（" "）	（double）quotation marks	引号
' '（''）	single quotation marks	单引号
——	dash	破折号
-	hyphen	连字号
'（'）	apostrophe	撇号，省字号
…	ellipsis（or suspension points）	省略号
（ ）	round brackets（or parentheses，curves）	（圆）括号
[ ]	square brackets	方括号
< >	angle brackets	角括号
{ }	braces	大括号
~	swung dash（or tilde）	代字号（或波浪号，波浪字符）
^	caret	脱字号
/	slash（or virgule，slant）	斜线号
\	backslash	反斜线
*	asterisk	星号
§	section（or numbered clause）	分节号
‖	parallels	平行号
→	arrow	箭号

2. Common Commercial Symbols（常用商业符号）（Tab. 2）

Common Commercial Symbols（常用商业符号）   **Tab. 2**

Symbol（符号）	English reading or name（英语读法或名称）	Chinese reading or name（汉语读法或名称）
@	at, each	单价，电子邮箱符号，钢筋间距
%	percent	百分之…
‰	per thousand	千分之…
#	① number（before a number）:as，track #3 ② pounds（after a number）:as，5#	① …号（在数字前表示数目，如 track #3 即 3 号轨道） ② …磅（在数字后表示磅，如 5# 即 5 磅）

Continued

Symbol （符号）	English reading or name （英语读法或名称）	Chinese reading or name （汉语读法或名称）
¥	Renminbi yuan	（人民币）元
£	pound(s) sterling	（英）镑
$	dollar(s)	美元
®	registered trademark	注册商标
©	copyrighted	版权所有

3. The Greek Alphabet（希腊字母表）（Tab. 3）

The Greek Alphabet（希腊字母表）　　　　**Tab. 3**

Letter（字母）	Name（名称）	Pronunciation（读音）
α	alpha	['ælfə]
β	beta	['bi:tə, 'beitə]
γ	gamma	['gæmə]
δ	delta	['deltə]
ε	epsilon	[ep'sailən, 'epsilən]
ζ	zeta	['zi:tə]
η	eta	['i:tə, 'eitə]
θ	theta	['θi:tə]
λ	lambda	['læmdə]
μ	mu	[mju:]
ξ	xi	[gzai, ksai, zai]
π	pi	[pai]
ρ	rho	[rəu]
σ	sigma	['sigmə]
τ	tau	[tɔ:]
υ	upsilon	[ju:p'sailən, 'ju:psilən]
φ	phi	[fai]
ψ	psi	[psai]
ω	omega	['əumigə]

4. Fractions and Decimal Fractions（分数和十进制小数）

Let us first turn our attention to fractions. You have surely met the expressions "half" and "quarter", they are used when the denominator（分母）of the fraction equals 2 or 4. $\frac{1}{3}$ is read "one third". Other fractions are read in the same way. Thus we read $\frac{1}{5}$, $\frac{1}{6}$, $\frac{1}{7}$, $\frac{1}{10}$, $\frac{1}{25}$, $\frac{1}{100}$ as one fifth, one sixth, one seventh, one tenth, one twenty-fifth and one hundredth. These expressions are regarded as nouns and may therefore have a plural. Thus we read $\frac{2}{3}$

as two thirds; similarly $\frac{5}{6}$, $\frac{9}{10}$, $\frac{5}{100}$ are read as five sixths, nine tenths and five hundredths. However, if the last digit of the denominator is 1 or 2, then we do not read the fraction in the above-mentioned way. For example, we pronounce $\frac{5}{21}$ as "five over twenty-one". This method is also used in the other case. If the fraction is not a common one (e.g., $\frac{1}{1089}$ or $\frac{501}{1205}$), then we say "one over a thousand and eighty-nine" or "five hundred and one over twelve hundred and five".

Next, let us examine decimal fractions. They are very simple to pronounce. You just read the integral part of the number in the ordinary way, then say "point" (stands for "decimal point") and then read the decimal place one after the other. Thus 12.65 is read twelve-point-six-five; π correct to 6 decimal place, equals three-point-one-four-one-five-nine-two, correct to five significant figures (有效数字), equals three-point-one-four-one-six. When the decimal fraction is smaller than one, it is not usual in England to write, for instance, 0.56, but only .56. .56 is read "point-five-six", and .0007 is read "point-naught-naught-naught-seven" or more usually "point-three 0's-seven".

Now for algebraical expressions (代数表达式), fractions are again read "over". $\frac{2a-1}{ax+b}$ is read $2a-1$ over $ax+b$ and brackets are indicated by the word "into", e.g., $(a+b)(a-b)$ is read "a plus b into a minus b". Powers are indicated by indices or exponents (指数). The index 2 is read "squared" and the index 3 "cubed" or "to the third". Other indices are read "to the fourth, to the fifth, to the minus second, to the nth". The identity (恒等式)

$$a^3 + b^3 = (a+b)(a^2 - ab + b^2)$$

reads "a cubed plus b cubed equals a plus b into a squared minus ab plus b squared". Or the equation

$$x^{-\frac{2}{3}} + \sqrt[5]{a^2} = 0$$

reads "x to the minus two thirds plus the fifth root of a squared equals zero".

5. BasicMathematical Operation Symbols and Formulas (基本运算符号与算式)

首先指出：表示数或量的字母，特别是出现在算式中的字母(字母变量通常以斜体字出现，已知其值的字母常数通常以正体形式出现)均按该字母的名称读。例如，英文字母 A(或 a)和希腊字母 π 分别按国际音标[ei]和[pai]读。

5.1 Arithmetic Operations (四则运算)

5.1.1 Addition (加法)

加号"+"读作 plus。例如，$1+2$ 读作 "one plus two"; $a+3$ 读作"a plus three"。

注：当"+"看作正号时，可读作"plus 或 positive"，如 $+5$ 读作"positive five"或"plus five"。

5.1.2 Subtraction (减法)

减号"−"读作 minus。例如，$3-2$ 读作 "three minus two"; $5-b$ 读作"five minus b"。

注：当"−"看作负号时，可读作"minus 或 negative"，如 -5 读作"negative five"或"minus five"。

5.1.3 Multiplication (乘法)

乘号"×" " · "或省略均读作"times"或"multiplied by"。例如，$3×2$ 读作"three times two"或"three multiplied by two"; $a \cdot b$ 或 ab 都读作："a times b"或"a multiplied by b"。

5.1.4 Division (除法)

除号"÷"读作"divided by"; 用"/"或"—"表示除法时读作"over"或"divided by"。例如，$10÷4$ 读作"ten over four"或"ten divided by four"; a/b 或 $\frac{a}{b}$ 读作"a over b"或"a divided by b"。

注：分数(式)的读法比较复杂，把 $\frac{a}{b}$ 看作分数(式)时通常读作"a over b"，但有许多具体情形采用其他读法，如：$\frac{1}{4}$ 读作"a quarter"; $\frac{3}{4}$ 读作"three quarters"等。

5.1.5 Proportion (比例)

比例符号"："读作"is to"。例如，$a:b$ 读作"a is to b"; 有时读作"the ratio of a to b"。

5.1.6 Power（乘方）或 Exponent（指数）

x 的 n 次方 x^n 一般读作"x to the nth power"或"x to the power of n"。a^b 读作"a to the bth power"或"a to the power of b"。例如，ab^n 读作"a times b to the nth power"或"a times b to the powe of n"；b^{n-1} 读作"b to the n minus one power"或"b to the power of n minus one"；b^{m+n} 读作"b to the m plus n power"或"b to the power of m plus n"。但是，当 $n=2,3$ 时读法不同：x^2 读作"x squared"或"x square"；x^3 读作"x cubed"或"x cube"。

5.1.7 Extraction（开方）

x 的 n 次方根 $\sqrt[n]{x}$（或 $x^{\frac{1}{n}}$）一般读作"the nth root of x"。例如，$\sqrt[5]{a^2}$ 读作"the fifth root of a squared"。当 $n=2,3$ 时读法不同：$\sqrt{x}$（或 $x^{\frac{1}{2}}$）读作"the square root of x"，$\sqrt[3]{x}$（或 $x^{\frac{1}{3}}$）读作"the cube root of x"。

5.2 Logarithm（对数）

"$\log_a b=m$" is read as "the logarithm of b to the base a equals m".

5.3 Function and Its Derivative（函数及其导数）

The symbol $f(x)$ is read as "f of x". The symbol $f'(x)$ is read as "f prime of x" [$f'(x)$ 读成 f 一撇 x]. The symbol $f^{(n)}(x)$ or $f^{(n)}$ is read as "the nth derivative of $f(x)$ or the nth derivative of f". Symbols "$'$" and "$''$" are read as "prime" and "double prime" respectively.

5.4 Comparison of Quantities（大小关系）

等于：$a=b$ 读作"a is equal to b"或"a equals b"，也可读作"a is b"；$x+2=7$ 读作"x plus 2 is equal to 7"或"x plus 2 equals 7"，也可读作"x plus 2 is 7"。

不等于：$a \neq b$ 读作"a is not equal to b"或"a does not equal b"，也可读作"a is not b"。

恒等于：$a \equiv b$ 读作"a is identical to b"或"a is identical with b"。

近似等于：$a \approx b$ 读作"a is approximately equal to b"。$\pi \approx 3.14$ 读作"π is approximately equal to three point one four"。

小于：$a < b$ 读作"a is less than b"；$3+a < m-3$ 读作"3 plus a is less than m minus 3"。

大于：$a > b$ 读作"a is greater than b"。

小于或等于：$a \leqslant b$ 读作"a is less than or equal to b"。

大于或等于：$a \geqslant b$ 读作"a is greater than or equal to b"。

5.5 Signs of Grouping（括号）

圆括号（）、方括号[ ]、大括号{ }分别读作"parenthesis"（或"round brackets"）"square brackets""braces"。

括号的完整读法比较麻烦，如$(a+b)$，先读左半括号（open parenthesis），再读 $a+b$，最后读右半括号（close parenthesis），整个式子读作"open parenthesis a plus b close parenthesis"。为了简便，$(a+b)$ 可改读成"the quantity a plus b"。圆括号换成方括号、大括号时的情形类似。

更进一步的简化是省去 the quantity，如$(a+b)(a-b)$ 读作"a plus b into a minus b"。这里 into 代表"乘"。

5.6 Special Evaluation（特殊求值）

绝对值（Absolute value）：$|x|$ 读作"the absolute value of x"。

最大值（Maximum value）：$\max f(x)$ 读作"the maximum value of $f(x)$"，$\max\{x_1, \cdots, x_n\}$ 读作"the maximum value of the series x sub one to sub n"。

最小值（Minimum value）：$\min f(x)$ 读作"the minimum value of $f(x)$"，$\min\{x_1, \cdots, x_n\}$ 读作"the minimum value of the series x sub one to sub n"。

求累加和（the sum of the terms indicated）：

$$\sum_{k=1}^{n} a_k$$

读作"Capital sigma a_k from k equals one to k equals n"或"Sum of all a_k from k equals one to n"；

$$\sum_{k=1}^{\infty} a_k$$

读作"Capital sigma a_k from k equals one to k equals infinity"或"Sum of all a_k from k equals one to infinity"。

求连乘积（the product of the terms indicated）：$n!$ 读作"Factorial n";

$$\prod_{k=1}^{n} b_k$$

读作"Product of all b_k from k equals one to n";

$$\prod_{k=1}^{\infty} b_k$$

读作"Product of all b_k from k equals one to infinity"。The symbols $\sum$ and $\prod$ also read "capital sigma and capital pi" respectively.

Appendix 2: Commonly Used English-Chinese Professional Terms for Civil and Architectural Engineering
（附录 2：常用土木建筑类专业词汇英汉对照）

abhesive [əb'hiːsiv] n. 阻黏剂

abnormal end [计] 异常终止

abrasive floor 防滑地板

Abromovitz [ˌæbrə'məuviz] 阿布拉莫维茨(建筑师)

abscissa [æb'sisə] n. 横坐标

absolute coordinate 绝对坐标

absolute value 绝对值

abutment [ə'bʌtmənt] n. 桥墩，支座

accelerated cement 快凝水泥

accelerator [ək'seləreitə(r)] n. 促凝剂，速凝剂

acceptance of hidden subsurface work 隐蔽工程验收

acceptance of sub-divisional work 分项工程验收

acceptance of tender 得标

access ['ækses] n. 通道，进入，(对计算机存储器的)访问；~ control 访问控制；~ permission 访问许可；~ eye 清扫孔，检查孔；~ hole 检修孔；~ plate 检修孔盖板

accessibility [əkˌsesə'biliti] n. 可达性

accident (fatality) rate 事故(死亡)率

accordion shades 折叠式活动隔断，屏风

accretion [ə'kriːʃ(ə)n] n. 冲积层

acid ['æsid] n. 酸

ACM (Association for Computer Machinery) 计算机协会

acoustical insulation 隔声

active desktop 活动桌面

active earth pressure 主动土压力

active window 活动窗口

acute angle 锐角

acute triangle 锐角三角形

adapter card 适配卡

adhesive bitumen primer 冷底子油

administration of the construction contract 施工合同管理

administrator [əd'ministreitə(r)] n. 管理员

admixture [əd'mikstʃə(r)] n. 外加剂

adobe [ə'dəubi, ə'dəub] n. 砖坯，土砖

advance signing 预告标志

advances of starting 开工预付款；progress payment 工程进度款

aerated concrete 加气混凝土

aerial ladder 消防梯

aerial photogrammetry 航空摄影测量

aesthetics [iːs'θetiks] n. 美学，美感，审美观

age of concrete 混凝土龄期

age hardening 时效硬化

aggregate ['ægrigeit] n. 骨料

agriculture and animal husbandry building 农牧业建筑

air void ratio(土)空隙比

air-conditioning system 空调系统

air-entrained concrete, air-entraining concrete, aerated concrete 加气混凝土

air-entrained agent 加气剂

air-raid defense 人防

aisle [ail] n. 过道，走廊，通道

albery ['ælbəri] n. 壁橱，壁龛

Albrecht Durer ['ælbret 'djuə] 阿尔布雷德·丢勒(德国文艺复兴时期画家、版画家)

alée or allee [ə'liː] n. 林荫宽步道，林荫小径

algebra ['ældʒibrə] n. 代数

algebraical expression 代数表达式

algorithm ['ælgəriðəm] n. 算法，运算法则

Alistair Bevington ['æliˌstɛə 'beviŋtən] 阿利斯泰尔·贝文顿(建筑师)

alkali ['ælkəlai] n. 碱

alkaline ['ælkəlain] adj. 碱性的；n. 碱性

allowable (or permissible, or tolerable) bearing capacity 容许承载力

allowable stress approach 容许应力法

alloy ['æləi] n. 合金

almery ['ælməri] n. 壁橱，贮藏室

aluminum alloy 铝合金

American Institute of Architects (AIA)美国建筑师学会

analytic geometry 解析几何

analytil method 解析法

anchor ['ænkə] n. 锚；v. 固定，栓住

anchorage ['ænkəridʒ] n. 锚固

anchored sheet pile wall 锚定板桩墙

angles ['æŋglz] n. 角钢

annealed copper 退火铜；韧铜

ANSl (American National Standard Institute)美国国家标准协会

anti-seismic joint 防震缝

anti-symmetry 反对称(性)

Appendix 2: Commonly Used English-Chinese Professional Terms for Civil and Architectural Engineering
（附录2：常用土木建筑类专业词汇英汉对照）

235

anti-virus program 防病毒程序

antivirus software 杀毒软件

apartment [ə'pɑ:tmənt] n. 公寓住宅，单元住宅

application program 应用程序

approach zero 趋于零

aqueduct ['ækwidʌkt] n. 水管，高架渠，渡槽

arborvitae [ˌɑ:bə'vaiti:] n. 侧柏

arc [ɑ:k] n. 弧

arc length 弧长

arc welding 电弧焊

arcadian architecture, pastoral architecture 田园式建筑

arch structure 拱结构

architect ['ɑ:kitekt] n. 建筑师

architectural [ˌɑ:ki'tektʃərəl] adj. 建筑上的，建筑学的；~ design institute 建筑设计院；~ design theory 建筑设计理论；~ elevation view (or drawing) 建筑立面图；~ (or building) mechanics 建筑力学；~ plan view (or drawing) 建筑平面图；~ section view (or drawing) 建筑剖面图；~ working drawing 建筑施工图；~ appearance 建筑外观；~ area 建筑面积；~ lighting 建筑采光，建筑照明；~ perspective 建筑透视图；~ specifications, building code 建筑规范

Architectural Graphic Standards（GB/T 50104—2010）《建筑制图标准》GB/T 50104—2010

architecture ['ɑ:kitektʃə] n. 建筑学；~ sketch 建筑草图

arcology [ɑ:'kɔlədʒi] n. 建筑生态学

arena [ə'ri:nə] n. （古罗马圆形竞技场内的）斗技场，表演场；sports ~s 运动场

arithmetic [ə'riθmətik] adj. 算术的；n. 算术

arrow keys 箭头键，方向键

Art Court，Yi Court 艺苑

art design，artistic designing 艺术设计

art of architecture 建筑艺术

artificial daylight 人工采光

artificial illumination 人工照明

artist-craftsman 艺术家兼手艺人

ASCI (American Standard Code for Information Interchange) 美国信息交换标准代码

ashlar ['æʃlə] n. 方石堆，装饰屋内墙面的石板

ASP (Application Service Provider) 应用服务提供商

asphalt ['æfælt] n. 沥青，柏油

assembler [ə'semblə(r)] n. 汇编程序，装配工

ASTM (American Society for Testing Materials) 美国材料试验协会

asymptote [英 'æsimptəut，美 'æsim,təut] n. 渐近线

at-grade intersection (or crossing) 平面交叉

attached dwelling (or house) 毗连住宅

auger ['ɔ:gə] n. 螺旋钻；~ boring 螺旋钻探

axial ['æksiəl] adj. 轴的，轴向的

auxiliary storage [计] 辅助(或外部)存储器

axis of symmetry 对称轴

azure ['æʒə(r)，'æzjʊə(r)] dragon 青龙

B2B (Business to Business) 商业机构对商业机构的电子商务

B2C (Business to Consumer) 商业机构对消费者的电子商务

Babylonia [ˌbæbi'ləunjə] n. 巴比伦王国

back view 背立面

back-acting shovel 反向铲

bail of bid 投标保证金

balance (or equilibrium) equation 平衡方程

balcony ['bælkəni] n. 阳台

ballast ['bæləst] n. 石渣，碎石

balustrade [ˌbælə'streid], handrail ['hændreil] n. 栏杆，扶手

bamboo scaffolding 竹脚手架

band iron 扁铁，扁钢

bandwidth ['bændwidθ] n. 带宽

bar [bɑ:] n. 棒，条，杆件，（粗）钢筋；~ cutter 钢筋切断机；~ list 钢筋表；~ spacing 钢筋间距

Baroque architecture 巴洛克建筑

barrel ['bærəl] n. 岩芯管，筒状物

barrier-free design 无障碍设计

barycenter ['bæri,sentə(r)] n. 重心，质心

base board 踢脚板

basic module 基本模数

batching ['bætʃiŋ] n. 配料

Bauhaus ['bauhaus] n. 包豪斯(建筑学派)，现代主义建筑

bay [bei] n. 湾，跨；~ window 飘窗

beam [biːm] n. 梁；～and slab structure 梁板结构；～with fixed ends 固端梁

beam-and-column construction 梁柱结构（框架结构）

beam-and-girder construction 主次梁梁格结构

bearing capacity 承载力；～of a pile 单桩承载；～of foundations 地基承载力

bearing wall 承重墙

bed joint 底层接缝，平缝，平层节理

Beijing lacquer carving 北京漆雕

belled pier 扩底墩

bend [bend] v. n. 弯曲；～bar（or steel），bent reinforcement bar，bent-up bar 弯筋

bending moment 弯矩；～diagram 弯矩图；～envelope 弯矩包络线

bending stiffness 弯曲刚度

bending strength 抗弯强度

bending stress 弯曲应力

biaxial [baiˈæksiəl] adj. 二轴的

bid call 招标

bid opening 开标

bid price 出价，投标价格

bidder [ˈbidə] n. 投标人

bidding sheet 标价单

biennial architecture exhibition 建筑双年展

binding agent 胶粘剂

binding reinforcement 绑扎钢筋

BIOS（Basic Input/Output System）基本输入输出系统（BIOS 是电脑启动时加载的第一个软件）

bird's eye view 鸟瞰图

Bird's Nest，The National Stadium 鸟巢

bitumen [ˈbitjumin] n. 沥青

bituminous dampproofing 沥青防潮层

black-top，blacktop n. ［公路］沥青路面；柏油路

blackwood furniture 红木家具

bleeding [ˈbliːdiŋ] n. 泌水

blind window，window shutter 百叶窗

blinding concrete 基础（地基）垫层混凝土

block resonant test 块体共振试验

blocking course，parapet（wall）女儿墙

blockwork [ˈblɔkwəːk] n. 砌块工程，砌块墙体

Blue Waves（or Cang Lang）Garden 沧浪亭

bluetooth headset 蓝牙耳机

bolt [bəult] n. v. 螺栓（杆），插销，用螺栓固定

bolting，bolt connection 栓接，螺栓连接

bond [bɔnd] n. 黏结，结合力，结合物，砌合；～beam 结合梁，圈梁；～stone 砌合石

bookbinding 书籍装帧设计

Boolean logic 布尔逻辑

Boolean operation 布尔运算

boot up ［计］启动

bore [bɔː] n. 孔，枪膛；v. 钻孔

borehole [ˈbɔːhəul] n. 钻孔

boring [ˈbɔːriŋ] n. 钻探；～machine 钻机

bottleneck section 瓶颈路段

bottom chord，lower chord 下弦杆

boulder [ˈbəuldə] n. 漂石

boundary [ˈbaundəri] n. 边界，分界线；～lines of roads 道路红线

boundary condition 边界条件

box caisson 箱形沉箱

box foundation 箱形基础

brachy-axis [ˈbræki ˈæksis] 短轴

bracing [ˈbreisiŋ] n. 拉条，撑杆，支撑

brick lintel 砖砌过梁

brickwork [ˈbrikwəːk] n. 砖砌，砖砌的建筑物

bridgepier [ˈbridʒˈpiə] n. 桥墩

brittle [ˈbritl] adj. 脆性的，易损坏的；～failure 脆性破坏；～fracture 脆裂

BRKT＝bracket 牛腿

broadband network 宽带网络

bronze craft 青铜工艺

bronze mirror 铜镜

browser [ˈbrauzə(r)] n. 浏览器

Brundtland Commission's report 布伦特兰报告

BSMT＝basement 地下室

buckle [ˈbʌkl] n. 屈曲，压曲；v. 弄弯，皱曲，翘曲

buckling [ˈbʌkliŋ] n. 压曲

buggy [ˈbʌgi] n. 小推车

building [ˈbildiŋ] n. 建筑物，房屋；～density，density of ～建筑密度；～line，～setback line 建筑收进线，建筑红线；～rubble（or debris）建筑垃圾；～equipment 建筑设备；～height 建筑高度；～orientation 建筑朝向；～code（BC）建筑法规，建筑规范；

Appendix 2: Commonly Used English-Chinese Professional Terms for Civil and Architectural Engineering
（附录2：常用土木建筑类专业词汇英汉对照）

237

~ industrialization 建筑工业化；~ law 建筑法；~ storm sewer 房屋雨水管；~ module 建筑模数；~ permits for construction 建筑施工执照；~ physics 建筑物理

built-in ['bilt'in] adj. 固定的，嵌入的，内装的，固有的；~ cupboard 壁橱；~ fitting 预埋件

built-up section 组合断面；组合型材

bulk storage 大容量存储器

bundled-tube structure 成束筒结构

burnt clay 黏土砖

buttress ['bʌtris] n. 拱壁，扶壁，支持物

Byzantine [英 bai'zæntain，美 'bizənti:n] architecture 拜占庭式建筑

cable ['keibl] n. 索，缆；~ structure，~-supported construction 悬索结构；~-stayed bridge 斜拉桥

cadastral [kə'dæstrəl] adj. 地籍的，地产的

caisson ['keisən] n. 沉箱

calculation [,kælkju'leiʃn] n. 计算

calculus ['kælkjuləs] n. [数] 微积分，微积分学（differential and integral calculus 的简写）

campus area network 校园网

canal [kə'næl] n. 沟渠，运河，水道

canopy ['kænəpi] n. 雨篷，天篷

cantilever ['kæntili:və] n. 悬臂，伸臂；~ beam 悬臂梁；~ed balcony 悬挑阳台，凸阳台

capillary [kə'piləri] n. 毛细血管，毛细管现象

capital of Gothic column 哥德式柱头

carbon disulphide 二硫化碳

carport ['ka:pɔ:t] n. 汽车棚

carry-over factor 传递系数

Cartesian [ka:'ti:ziən] adj. 笛卡尔的；~ coordinate system 笛卡尔坐标系；~ geometry 解析几何，笛卡尔几何学

cartographic [,ka:tə'græfik] adj. 地图制图学的

cast [ka:st] v. 浇筑，浇捣；~-in-place（or ~-in-situ）concrete 现浇混凝土

caterpillar crane 履带式起重机

causeway ['kɔ:zwei] n. 长堤，堤道，铺道

cavity（or hollow）brick 空心砖

cavity（or hollow）wall 空心墙，空斗墙

cement [si'ment] n. 水泥；~ floor 水泥地面；~ mortar 水泥砂浆

cement mortar 水泥砂浆

census tract 人口普查区，普查地段

center to end, C to E 中心到端面的距离

center-to-center, C to C 中心距，中到中距离

Central Business District（CBD）中心商务区

central line 中心线，中线

central lobby 中央走廊，中央大厅

Central Processing Unit（CUP）中央处理器，中央处理单元

centroid ['sentrɔid] n. 形心，质心，重心

ceramic [si'ræmik] adj. 陶瓷的，制陶艺术的；n. 陶瓷，陶瓷制品；~ mosaic，~ mosaic tile 陶瓷锦砖，马赛克；~ tile 瓷砖

Certificate Authority（CA）证书认证

cesspit, digestion（or septic, or sewage）tank 化粪池

Chad Floyd [tʃæd flɔid] n. 查德·弗洛伊德（建筑师）

chain-pull switch 拉线开关

chalet [英 'ʃælei，美 ʃæ'lei] n. 避暑山庄，小木屋

channel ['tʃæn(ə)l] n. 沟渠，海峡，通道，槽钢

channelizing lines 渠化线

characteristic function 特征函数

Charles W. Moore [tʃa:lz 'dʌblju:mɔ:] n. 查尔斯·W·摩尔（建筑师）

chartered structural engineer 特许结构工程师，注册结构工程师

chasm ['kæzəm] n. 路堑，鸿沟，深坑

chicken-wire 铁丝织网，细号钢丝网

chief architect 总建筑师

chief engineer 总工程师

Chinese Imperial City Planning 中国皇城规划

chip [chip] n. 芯片；~ circuitry 芯片电路；~ set ['tʃipset] n. 芯片组

chord [kɔ:d] n. 弦

chromatics [krəu'mætiks] n. 色彩学

chromium ['krəumiəm] n. 铬

Churchill P. Lathrop Gallery 丘吉尔·P·莱斯罗普画廊

ciphertext ['saifə,tekst] n. [计] 密文

circumferential [sə,kʌmfə'renfəl] adj. 圆周的，环形的

circular arc 圆弧

circular cone 圆锥

City Beautiful Movement 城市美化运动

civil ['sivl] adj. 土木工程的，民用的；~ engineering 土木工程；~ architecture (or building)民用建筑

Clair County Regional Education Service Agency 克莱尔县地区教育服务机构

clamp [klæmp] v. n. 夹紧，夹子，夹钳，夹具

classic architecture 古典建筑

Claudia Thornton ['klɔːdiə 'θɔːntən] n. 克劳迪娅·桑顿(建筑师)

clay [klei] n. 黏土，泥土；~ brick 黏土砖；~ ey ['kleii] adj. 黏土的，涂了黏土的

clear span 净跨

clearance height 净空高度

clerestory ['kliəstɔːri] window 天窗

clevis ['klevis] n. U 形钩，V 形钩，马蹄钩，夹板

clipboard ['klipbɔːd] n. [计] 剪贴板

Cloisonné, cloisonne [ˌklɔizə'nei] n. 景泰蓝

close interval 闭区间

closely spaced perimeter tube 密布(小柱距)外框筒

closet ['klɔzit] n. 盥洗室，厕所，卫生间

clouded glass, ground glass 毛玻璃，磨砂玻璃

CMOS Setup Utility，CMOS 设置实用程序

CMOS (Complementary Metal Oxide Semiconductor)互补金属氧化物半导体，是指制造大规模集成电路芯片用的一种技术或用这种技术制造出来的芯片，是电脑主板上的一块可读写的用于保存 BIOS 设置程序设置完电脑硬件参数后的数据的 RAM 芯片(属于硬件)

CMYK 印刷色彩模式；C—Cyan 青色；M—Magenta 洋红色；Y—Yellow 黄色；K—Black 黑色

coarse aggregate 粗骨料

coarse sand 粗砂；medium sand 中砂；fine sand 细砂；mixed sand 混合砂

coated glass, glass curtain wall 玻璃幕墙

coaxial cable 同轴电缆

cobble ['kɔbl] n. 圆石，卵石；boulder 大卵石；pepple 小卵石；rubble 毛石；gravel 碎石

code [kəud] n. 规范；~ for architectural design 建筑设计规范

coefficient [kəui'fiʃənt] n. 系数；~ of permeability 渗透系数；~ of thermal expansion 热膨胀系数；~ of regression 回归系数

cohesive [kəu'hiːsiv] adj. 有黏聚力的，黏性的

cold-draw 冷拉，冷拔

collage [kə'laːʒ] n. 拼贴画

collapsible [kə'læpsəbəl] adj. 活动的，可分解的；~ (or slumping) loess 湿陷性黄土

collar tie beam 圈梁

color composition 色彩构成

Colosseum [ˌkɔlə'siəm] n. 古罗马圆形大剧场；colosseum 公共娱乐场

column ['kɔləm] n. 圆柱，列

combination beam 组合梁

combination construction 混合结构

commemorative architecture 纪念性建筑

commercial buildings 商业建筑物，商业房屋

commodity [kə'mɔdəti] n. 实用(适用)

compacted fill 压实填土，夯实填土

compacted soil 压实土

compaction by layers 分层填土夯实

compaction by rolling 碾压

compaction by vibration 振动压实

compartmentation 隔断

compatibility [kəmˌpætə'biliti] n. 相容性，协调性；~ equation (of strain)(变形)协调方程

compiler [kəm'pailə(r)] n. 编译器，汇编者

complementary [ˌkɔmpli'mentəri] adj. (互)余的

completion acceptance 竣工验收

completion date 竣工日期

compression [kəm'preʃən] n. 压缩；~ bar (or steel)受压钢筋

compressive [kəm'presiv] adj. 压缩的，有压缩力的；~ strength 抗压强度

computability [kəmˌpjuːtə'biləti] n. 可计算性

Computer Aided Architectural Design (CAAD) 计算机辅助建筑设计

Computer Aided Design (CAD)计算机辅助设计

Computer Aided Instruction (CAI)计算机辅助

Appendix 2: Commonly Used English-Chinese Professional Terms for Civil and Architectural Engineering
（附录 2：常用土木建筑类专业词汇英汉对照）

239

教学

Computer Aided Manufacturing（CAM）计算机辅助制造

computer malfunction 电脑故障

computer network technology 电脑网络技术

computer virus 计算机病毒

Computer-Aided Engineering（CAE）计算机辅助工程

computerize [kəm'pju:təraiz] vt. 用电脑处理

concave downward 凹向下的

concealed work 隐蔽工程

concentric [kən'sentrik] 轴心的，同心的

concept sketch 概念草图

concrete['kɔŋkri:t] n. 混凝土；~ cover 混凝土保护层；~ mix design 混凝土配合比设计；~ placement 混凝土浇筑；~ structure 混凝土结构；~ mixer 混凝土搅拌机

conductivity [ˌkɔndʌk'tiviti] n. 传导性

cone [kəun] n. 锥体，锥形漏斗；~ penetrometer 圆锥贯入仪

conservation of energy 能量守恒

conservation of historical and cultural city 历史文化名城保护

conservatory [kən'sə:vətri] n. 暖房

consolidation [kənˌsoli'deiʃən] n. 固结；~ settlement 固结沉降

constant ['kɔnstənt] n. 常数；~ of integration 积分常数

constraint [kən'streint] n. 约束，固定

construction [kən'strʌkʃən] n. 施工，建造；~ joint 施工缝；~ progress 施工进度；~ (or building) material 建筑材料；~ administration 施工管理；~ and installation work 建筑安装工程；~ company 建筑公司；~ economics 建筑经济；~ in process 在建工程；~ industry 建筑业；~ management plan 施工组织设计；~ organization 施工单位；~ period 施工工期；~ site 施工现场；~ al drawing 施工图

Construction Law of PRC 中华人民共和国建筑法

constructive detailed planning 修建性详规

continuous beam 连续梁

contraction [kən'trækʃən] n. 收缩

contractor [kən'træktə] n. 承包人，承包商，包工

control panel [计] 控制面板

convex ['kɔnveks] adj. 凸的，凸出的

conveyor [kən'veiə] n. 输送机

cooling fan 冷却风扇

coordinate [kəu'ɔ:dineit] n. 坐标；~ axis 坐标轴；~ system 坐标系

cordless mouse 无绳鼠标，无线鼠标

core [kɔ:n] n. 岩芯，芯样，核心；~ barrel 岩芯钻筒；~ supported structures 核心筒

core supported structures 核心筒

core-in-core (or tube-in-tube) structure 筒中筒结构

corridor ['kɔridɔ:(r)] n. 走廊

corrode [kə'rəud] v. 腐蚀，锈蚀

corrosive [kə'rəusiv] adj. 腐蚀的

corrugation [ˌkɔru'geiʃən] n. 起皱，波纹

cosine law 余弦定律

couple ['kʌpl] n. 对，力偶；v. 耦连，耦合

Couple Garden, Ou Garden 耦园

coupled shear walls and beams 联肢剪力墙-梁

course [kɔ:s] n. （砖）层，行

crack [kræk] n. 裂缝，裂纹；v. （使）破裂

cracking ['krækiŋ] n. 开裂，破（爆）裂

craftsman ['kra:ftsmən] n. 工匠

creep [kri:p] n. 徐变，蠕变

critical ['kritikəl] adj. 临界的，极限的；~ path method (CPM) 关键路径法；~ value 临界值

crookedness ['krukidnis] n. 弯曲，扭曲，歪斜

cross section 横断面，截面，断面图，横剖面图

cross wall 横墙

cruise [kru:z] vi. n. 勘查，巡航

crushing ['krʌʃiŋ] n. 压碎

crystallographic [ˌkristələu'græfik] adj. 晶体学的，结晶的，结晶学的

cube [kju:b] n. 立方体

culvert ['kʌlvət] n. 涵洞

curb [kə:b] n. 路缘(石)，道牙；v. 设路缘石

cure [kjuə] v. 养护

curing ['kjuəriŋ] n. 养护

curtain wall 幕墙

curvature ['kə:vətʃə] n. 曲率，弯曲

curvilinear equation 曲线方程

curved roof 弧形屋顶

cut, copy and paste［计］剪切、复制、粘贴

cut-and-try 试验性的

cyber-surroundings 网络环境

cylinder［'silində］n. 圆筒，圆筒状物

cylindrical［si'lindrikəl］adj. 圆柱体的

dam［dæm］n. 坝

damp-proof coating（or course）防潮层

dark room 暗室

Darmstadt［'dɑːmstæt］n. 达姆施塔特（德国城市）

database［'deitəbeis］n. 数据库，数据库技术

datum［'deitəm］n. 基准面，基准线，基准点，资料，数据；～plane 基准面

DBMS（DataBase Management System）数据库管理系统

DDR（High Definition Digital Recorder）高清晰度数字录像机

dead load 恒载，静载

debris［'debriː］flow 泥石流

debugger［,diː'bʌgə(r)］n. 调试程序

decimal fraction 十进制小数

decision trees and simulation models 决策树与仿真模型

deck［dek］n. 甲板

decoder［diː'kəudə(r)］n. 译码器

decryption［diː'kripʃn］n. 解密

deep foundation 深基础

deflection［di'flekʃən］n. 挠度，挠曲，偏差，倾斜

deformation［,diːfɔː'meiʃən］n. 变形，形变；～joint 变形缝；～compatibility condition 变形协调条件

degree of freedom 自由度

deice［diː'ais］v. 除冰，防冻

deicing［,diː'aisiŋ］salt 除冰（化冰）盐，融雪剂

delight［di'lait］n. 愉悦，美观，高兴

demodulation［'diː,mɔdju'leiʃn］n. 解调

demographic［,demə'græfik］map 人口地图

den［den］n. 书房，工作室

denominator［di'nɔmineitə(r)］n. 分母

density［'densiti］n. 密度

derivative［di'rivətiv］n. 导数

derrick［'derik］n. 人字起重机，摇臂吊杆

desiccation［,desi'keiʃn］n. 干燥，干缩，干裂

design［di'zain］v. n. 设计；～codes 设计规范；～speed 设计车速；～organization 设计单位；～scheme 设计方案

designed longevity［lɔn'dʒeviti］设计寿命

designer-craftsman 设计师兼工匠

Dessau［'disɔː］n. 德绍（德国城市）

detached dwelling（or house）独立式住宅

detail drawing 详图，大样图，细部图

detailed planning 详细规划

determinant［di'təːminənt］n.［数］行列式

determinate［di'təːminit］adj. 确定的，决定的；n.［数］行列式；～structure 静定结构

developable surface 可展曲面

development area 开发区

development organization 建设单位

device driver［计］设备驱动程序

diaphragm［'daiəfræm］n. 横隔板，横隔墙，横隔膜；～wall, continuous concrete wall 地下连续墙

difference of elevation between inside and outside 室内外高差

differential［,difə'renfəl］adj. 微分的，差别的；n. 微分；～equation 微分方程；～equation of first order 一阶微分方程；～equation of higher order 高阶微分方程

differentiate［,difə'renfieit］v. 求导数，求微分

differentiation［,difə,renfi'eiʃn］n. 微分（法）

digital［'didʒitl］adj. 数字的；～certificate 数字证书；～multimedia technology 数字多媒体技术；～signature 数字签名；～simulation 数字模拟

dimension［di'mefən］n. 尺寸；～less［di'menfənlis］adj. 无量纲的，没有单位的

dinette［dai'net］n. 厨房旁的小餐室，小吃饭间

disk cleanup［计］磁盘清理

displacement［dis'pleismənt］n. 位移，～method 位移法

display adapter［计］显示适配器

distance education 远程教育

distort［dis'tɔːt］vt. 扭曲；～ion［dis'tɔːʃən］n. 扭曲，变形

distribute［dis'tribu(ː)t］vt. 分布，散布，

Appendix 2: Commonly Used English-Chinese Professional Terms for Civil and Architectural Engineering
(附录 2：常用土木建筑类专业词汇英汉对照)

241

分配；~d load 分布荷载；~d database 分布式数据库；~d processing 分布式处理；~d system 分布式系统

distributing bars, distribution reinforcement, distribution reinforcement bar 分布钢筋

distribution factor 分配系数

disturbed samples 扰动样

division [di'viʃən] n. 除，除法，部分，部门，分割，区域；quotient ['kwəuʃənt] n. 商

DL=dead load 恒载，自重

document ['dɔkjumənt] n. 文档，文件，资料

dogleg stair, half turn 双折楼梯

domain [dəu'mein] n. 区域，[数]定义域

door window 落地窗

dormant window, dormer window, lucarne [lu:'kɑ:n] n. 老虎窗，屋顶(天)窗

dot pitch 点距

dot-matrix 点阵

double reinforced beam 双筋梁

double door 双扇门

downgrade ['daungreid] n. 下坡，降级

downspout, drain spout, fall (or leader) pipe, rain-water leader 雨水管，落水管

drag-and-drop editing [计] 拖放编辑

drainage ['dreinidʒ] n. 排水；~system 排水系统

drawing board 绘图板

drilled pier 钻孔桩

drip line 滴水线

drop panel 无梁楼盖托板

drop-down list box [计] 下拉式列表框

drop-down menu [计] 下拉式菜单

ductile ['dʌktail] adj. 易延展的，延性的，韧性的；~failure 延性破坏

ductility [dʌk'tiliti] n. 延性

ducts and pipes 管道

dunny ['dʌni] n. 厕所，盥洗室

durability [,djuərə'biliti] n. 耐久性

dynamic IP address [计] 动态 IP 地址

dynamic penetration test 动力触探试验

earth pressure 土压力

earthenware ['ə:θənwɛə] n. 陶器，土器

earthquake['ə:θkweik] n. 地震；~intensity 地震烈度；~load 地震荷载；~resistant design 抗震设计

earth-retaining structures 挡土结构

earthwork ['ə:θwə:k] n. 土方工程，土石方工程；~quantity 土方工程量

eave [i:v] n. 屋檐

eavesdrop ['i:vzdrɔp] v. 偷听，窃听；n. 屋檐水

eccentric [ik'sentrik] adj. 偏心的

eccentrically-loaded [ik'sentrikli 'ləudid] 承受偏心荷载的

eccentricity [,eksen'trisiti] n. 离心率，偏心率

ecclesiasticism [i,kli:zi'æsti,siz(ə)m] n. 教会主义，拘泥教规

eco-city, ecocity, ecological city, an eco-friendly city 生态城市

ecodesign [,ekə'disain] n. 生态设计

ecological footprint 生态足迹

eddy ['edi] n. 涡流，漩涡，逆流

edge lines 行车道边线

Edward Larrabee Barnes 爱德华·拉华比·巴恩斯(建筑师)

effective depth 有效深度

efflorescence [,eflə'resns] n. 盐析，泛碱，渗斑

elastic mechanics 弹性力学

elasticity [elæs'tisiti] n. 弹力，弹性

electric moped 电动轻便摩托车

electrode [i'lektrəud] n. 电极，焊极，电焊条

electronic commerce (E-commerce)电子商务

electronic distance measurement (EDM)电子测距仪

electronic mail (E-mail)电子邮件

elevation [eli'veiʃən] n. 标高；~view 立面图

elevator ['eliveitə(r)] n. 电梯；~shaft 电梯井

ellipse [i'lips] n. 椭圆

embankment [im'bæŋkmənt] n. 路堤，堤岸

embedded computer 嵌入式计算机

embroider [im'brɔidə; em'brɔidə] vt. 刺绣，镶边

embroidery [im'brɔidəri, em'brɔidəri] n. 刺绣

empirical [im'pirikəl] adj. 经验的；~formula 经验公式

emulsion [i'mʌlʃən] n. 乳状液，乳胶

enameled tile 琉璃瓦，釉面砖

encode［in'kəud］n. 编码

encryption and authentication system 加密认证系统

encryption key 加密密钥

end user 终端用户

engineering geology 工程地质

engineering mechanics 工程力学

engineering geological prospecting 工程地质勘探

entity modeling 实体建模

entrain［in'trein］vt. 使（空气）以气泡状存在于混凝土中

entryway［'entriwei］n. 玄关

envelope diagram of internal force 内力包络图

environmental［in,vairən'mentl］adj. 环境的；~ art and design 环境艺术设计；~ design 环境设计；~ impact assessment 环境影响评价；~ planning 环境规划

equality［i:'kwɔiti］n. 等式，相等

equal-leg angle 等边角钢

equation［i'kweiʃən］n. 方程，等式

equidistant［,i:kwiədistənt］adj. 等距的

equilibrium［i:kwi'libriəm］n. 平衡；~ equation 平衡方程

ergonomics, human engineering 人体工程学

escalator［'eskəleitə(r)］n. 自动扶梯

Euler's formula 欧拉公式

evacuate［i'vækjueit］vt. vi. 疏散，撤离，排泄

evacuation［i,vækju'eiʃn］n. 疏散

evolution［,i:və'lu:ʃən］n. 进展，演变，［数］开方

excavation［,eskə'veiʃən］n. 挖土，挖掘

expanded（or shrinkage, or temperature）joint 温度缝，伸缩缝

expansion slot 扩展插槽

expert system 专家系统

expiry［iks'paiəri］n. 满期，到期，终结，死亡；~ date, deadline n. 截止日期

exponent［ik'spəunənt］n. ［数］指数

exponential function 指数函数

exponentiation［,ekspə,nenʃi'eiʃən］n. 幂运算

expressway［iks'preswei］n. 快速干道（部分立交），快速路

extension［ik'stenʃn］n. ［计］扩展名

exterior diagonal tube 外部斜撑筒

extraction［ik'strækʃn］n. ［数］开方

extreme value 极值

face shell（空心块材的）外壁

facility［fə'siliti］n. 设施，设备

factorial［fæk'tɔ:riəl］n. ［数］阶乘；adj. 阶乘的

factory building 厂房

familism［'fæmilizəm］n. 家庭主义

family or rumpus room 娱乐室

fatal error 致命错误

fatigue［fə'ti:g］n. 疲劳；vt. 使疲劳；vi. 疲劳

FCFS（First Come First Service）先到先服务

feasibility study 可行性研究

feed back vt. 反馈，回馈

feedback［'fi:dbæk］n. 反馈，回馈

Fellow of the American Institute of Architects（FAIA）美国建筑师学会会员

felt［felt］, malthoid［'mælθɔid］n. 油毡

ferrous metal 黑色金属

fiber concrete 纤维混凝土

fiber-optic cable 光纤

field erection 现场安装

field permeability test 现场渗透试验

FIFO（First In First Out）先进先出

figured（or patterned）glass 压花玻璃

file［fail］n. 文件

filler wall 填充墙

filling agency 填表单位（如专业行政机构或代理机构）

fine［fain］n. 细粒；~ aggregate 细骨料

fingerprint scanner 指纹扫描仪

finish［'finiʃ］v. 装修；~ ing material 装修材料

finite element method 有限单元法

finite slice method 条分法

finite-difference method 有限差分法

fire compartment 防火分区

fire rating 耐火等级

fine brick 耐火砖

fire-resistive material 耐火材料

firewall［'faiəwɔ:l］n. 防火墙

firmness［'fə:mnis］n. 坚固，安全

first-order polynomial 一阶多项式

fissure［'fiʃə］n. 裂缝；v. （使）裂开，（使）

Appendix 2: Commonly Used English-Chinese Professional Terms for Civil and Architectural Engineering
(附录 2：常用土木建筑类专业词汇英汉对照)

243

分裂

fixed end moment 固端弯矩

fixed window 固定窗

flange [flændʒ] n. 翼缘

flash memory 闪存

flash point, fire point 闪点，着火点(燃点)

flashing ['flæʃiŋ] n. 泛水

flat [flæt] adj. 平的，平坦的，扁平的；~ bar 扁钢；~ roof 平屋顶；~ slab and columns 平板-柱；~ slab and shear wall 平板-剪力墙；~ slab, shear walls and column 平板-剪力墙-柱；~ skylight 平天窗

flexibility [ˌfleksə'biliti] n. 挠度，柔度，柔韧性，机动性，适应性；~ coefficient 柔度系数

flexible ['fleksəbl] adj. 柔性的，可(易)弯曲的，软的；~ foundation 柔性基础

flexural ['flekʃərəl] adj. 弯曲的，挠曲的；~ rigidity 弯曲刚度

flocculate ['flɔkjuleit] v. 絮凝，(使)絮凝

flood [flʌd] n. 洪水，涨潮；~ runoff 洪水径流，洪峰流量

floor slab 楼板

floor load 楼面荷载

floor plan 楼层平面图

floor-to-ceiling height 楼面至顶棚高度，室内净高

floor-to-floor height 楼面至楼面高度(层高)

fluid mechanics 流体力学

fluidity [flu:'iditi] n. 流动性

foamed glass 泡沫玻璃

folded plate 折板

folder ['fəuldə] n. 文件夹

foliage ['fəuliidʒ] n. 观叶植物，叶子

font format 字样格式

footbridge ['futbridʒ] n. 人行桥，人行天桥

footnote ['futnəut] n. 脚注

force or flexibility method 力法或柔度法

form [fɔ:m] n. 模板

formability [fɔ:mə'biliti] n. 成形性

format ['fɔ:mæt] n. vt. 格式化，使格式化

formed steel, shape(d) steel 型钢

forming ['fɔ:miŋ] n. 成型，支模板；~ system 模板体系；stripping of the ~ 拆模

form-resistant structure 形抵抗结构

formula ['fɔ:mjulə] n. 公式

formulate ['fɔ:mjuleit] vt. 用公式表示

formwork ['fɔ:mwə:k] n. 模板(工程)

foundation [faun'deiʃ(ə)n] n. 基础；~ beam 基础梁；~ depth 基础埋深；~ pressure 基底压力；~ treatment 地基处理；~-bed 基础垫层

foundry ['faundri] n. 铸造，铸件，铸造厂

Fourier analysis 傅里叶分析

fractal geometry 分形几何

fraction ['frækʃn] n. 分数

fracture mechanics 断裂力学

frame [freim] n. 框(构，骨)架；~ structure；~ d structure 框架结构；~ tube structure 框筒结构；~-shear wall structure 框剪结构；~ work ['freimwə:k] n. 骨架，框架

Frank Gehry [fræŋk 'geri] n. 弗兰克·盖里(建筑大师)

Frank Stella [fræŋk 'stelə] n. 弗兰克·斯特拉(美国极简抽象主义画家)

Frankfurt ['fræŋkfət] n. 法兰克福(德国城市)

free-body diagram 自由体受力图，隔离体图

freeway ['fri:wei] n. 高速公路

fresh concrete 新浇混凝土，未硬化混凝土

friction coefficient 摩擦系数

friction pile 摩擦桩

function ['fʌŋkʃən] n. 函数；~ idea 函数思想；~ key 功能键；~ al relation 函数关系；~ al value 函数值

fuzzy mathematics 模糊数学

fuzzy-set theory 模糊集理论

gable, gable wall 山墙

Gallup poll ['gæləp pəul] 盖洛普民意测验

galvanized iron 镀锌铁皮，白铁皮，马口铁

gantry ['gæntri] n. 桶架，构台，起重机架

garage ['gærɑ:ʒ] n. 车库

gardenia [gɑ:'di:niə] n. 栀子花

gas-foaming admixture 发泡剂

gateway ['geitwei] n. [计]网关

general contractor 总承包商

general arrangement (or layout) drawing 总体布置图，总平面图

geodesy [dʒi'ɔdisi] n. 大地测量(学)

geodetic [dʒi:əu'detik] adj. 大地测量(学)的

Geographic Information System (GIS) 地理信

息系统

geography［dʒiˈɔgrəfi］n. 地理学，地形，地势

geology［dʒiˈɔlədʒi］n. 地质，地质学

geomancy［ˈdʒiːəumænsi］n. 风水，堪舆

geophysics［ˌdʒiːəuˈfiziks］n. 地球物理学

Georg Muche［dʒɔːdʒ mʌk］n. 乔治·蒙克（画家、编织艺术家和建筑师）

geotextile［ˈdʒioutekstail］n. 土工织物土工布

Gerhard Marcks［ˈgəːhəd maːks］n. 格哈德·玛克斯（雕塑家和陶艺家）

gigabit［ˈgigəbit］network 千兆网

gin pole 起重桅杆；~ derrick 桅杆起重机

girt［gəːt］n. 连系梁，墙梁，圈梁

glass curtain wall 玻璃幕墙

glass fiber reinforced plastics 玻璃纤维增强塑料，玻璃钢

Glenleven［ˌglenˈliːvən］Littleleaf linden 格伦利文小叶菩提树

Glenn［glen］Arbonies［ˈaːbəniəs］n. 格伦·阿伯尼斯（建筑师）

global system for mobile communication（GSM）全球移动通讯系统

glued board 胶合板

Gothic［ˈgɔθik］adj. 哥特式的；~ capital 哥德式柱头；~ Cathedral 哥特式教堂；~ house（or architecture）哥特式建筑

GPS（Global Positioning System）全球定位系统

GPU（Graphics Processing Unit）图形处理器

grade of memberships 隶属度

grade separation 立体交叉

gradient［greidiənt］n. 梯度

Grand Duke of Saxe-Weimar 萨克森-魏玛大公爵

granite［ˈgrænit］n. 花岗石，花岗岩

granite porphyry 花岗斑岩

Granite State 花岗岩之州，新罕布什尔州别名

graphic design 平面设计

graphical method 图解法

grassland［ˈgraːslænd］n. 牧场，草地，草原

gravel［ˈgræv(ə)l］n. 砾石；~ cobble；pebble ~（or stone）卵石

gravity load 重力荷载

green architecture（or building）绿色建筑

greenbelt［ˈgriːnbelt］n. 绿带

greenhouse［ˈgriːnhaus］, conservatory［kənˈsəːvətri］n. 温室

greenhouse effect 温室效应

greenware［ˈgriːnwɛə］n. 陶坯，未烧陶器

grey brick 青砖

grid structure 网架结构

grind［graind］v. 碾，磨，抛光；~ into a powder 碾成粉末

grit［grit］n. 砂粒，粗砂

grotto［ˈgrɔtəu］n. 岩洞，岩穴

ground engineering 地基工程

ground floor plan 底层平面图

groundwater surface 地下水位

group pile effect 群桩效应

group pile efficiency 群桩效率

grout［graut］n. 水泥浆，灰浆；v. 用水泥浆填塞

Guangdong-Hong Kong-Macao Large Bay Area 粤港澳大湾区

guardrail［ˈgaːdreil］n. 护栏，栏杆

guide signs 导向标志

gusset plates 节点板

Guthrie［ˈgʌθri］Theater 格思里剧院

gutter［ˈgʌtə(r)］n. 明沟，天沟

gypsum［ˈdʒipsəm］n. 石膏

hacker［ˈhækə(r)］n. ［计］黑客

Hagen［ˈhaːgən］n. 哈根（德国城市）

halfpace（or stair）landing, landing（or stair）platform 楼梯平台

hallway［ˈhɔːlwei］n. 门厅，过道

hand clapping 鼓掌，拍手，掌声响起

hand rail, handrail 扶手，栏杆

hangingwall subsidence 采场顶板沉降

Hannes Meyer［hænsˈmeijə］n. 汉斯·迈耶（建筑师）

Hard Disk Drive（HDD）［计］硬盘驱动器

hardcopy［ˈhaːdˈkɔpi］n. 硬拷贝

harden［ˈhaːdn］vt. 使变硬，使坚强；vi. 变硬；~ed concrete 硬化混凝土

hardening［ˈhaːdniŋ］n. 硬化

hardware［ˈhaːdwɛə］n. ［计］硬件

hardwood［ˈhaːdwud］n. 硬木

harp design 竖琴式设计

Harrison［ˈhærisn］n. 哈里森（建筑师）

head（end）joint 端灰缝

header［ˈhedə］n. 顶砖

header and footer ［计］页眉和页脚

Appendix 2: Commonly Used English-Chinese Professional Terms for Civil and Architectural Engineering
（附录2：常用土木建筑类专业词汇英汉对照）

245

headroom clearance 净空

height/thickness ratio 高厚比

helix ['hiːliks] n. 螺旋，螺旋状物；adj. 螺旋线的

hemp thread 麻丝

hemp，hair 麻刀

Herbert Bayer ['həːbət 'beijə] n. 赫伯特·拜耶(建筑师)

hessian ['hesiən] n. 粗麻布，浸沥青的麻绳

hierarchical directory structure 层次目录结构

high strength concrete 高强混凝土

high-early-strength Portland cement 早强水泥，快硬水泥

higher mathematics 高等数学

higher-order polynomial 高阶多项式

high-level language 高级语言

high-lift 高压的，高扬程的

high-rise building，high-rise 高层建筑

high-rise hotel 高层旅馆，高层饭店

highway ['haiwei] n. 公路

hinge [hindʒ] n. 铰链，枢纽

Hinnerk Scheper ['hinnək 'ʃiːpə] n. 辛涅克·谢柏(建筑师)

hip [hip] n. 屋脊线

histogram ['histəgræm] n. 柱状图，直方图

hockey rink 冰球场

hollow unit masonry 空心块材砌体

hollow wall 空心墙体

home page [计] 主页

homogeneous [ˌhəuməu'dʒiːniəs] adj. 均质的

honeycomb ['hʌnikəum] n. 蜂窝，空洞，麻面

Hong Kong-Zhuhai-Macao Bridge 港珠澳大桥

Hooke's Law 胡克定律

hoop [huːp] n. 环，箍；~ reinforcement，stirrup 环筋，箍筋

Hopkins Center 霍普金斯中心

HOPSCA: Hotel, Office, Park, Shoppingmall, Conference, Apartment 城市综合体

horizontal [ˌhɔri'zɔntl] adj. 水平的，水平线的；~ thrust of an arch 拱的水平推力

horticulture ['hɔːtikʌltʃə] n. 园艺，园艺学

host computer [计] 主机

Humble Administrator's Garden 拙政园

humidity [hju'miditi] n. 湿度

hybrid structure 混合结构

hydrated lime 熟石灰，消石灰

hydration [hai'dreiʃn] n. 水化，水化作用

hydraulics [hai'drɔːliks] n. 水力学

hydrogeology [ˌhaidrəudʒi'ɔlədʒi] n. 水文地质学

hydrology [hai'drɔlədʒi] n. 水文学

hydrostatic [ˌhaidrəu'steitik] adj. 静水的

hydrostatics [ˌhaidrəu'stætiks] n. 流体静力学

hyperbola [hai'pəːbələ] n. [数] 双曲线

hyperbolic paraboloid surface 双曲抛物面

hyperlink ['haipəliŋk] n. 超链接

HyperText Transport Protocol (HTTP) 超文本传输协议

hypotenuse [hai'pɔtənjuːz] n. 三角形的斜边

icon ['aikɔn] n. 图标

ideal constraint 理想约束

identity [ai'dentiti] n. [数] 恒等式

ideological and political elements 思政元素

if and only if (有时省略成 iff) 或 when and only when 当且仅当

igneous rock 岩浆岩，火成岩

impact factor ['impækt'fæktə(r)] 冲击系数

imaginary (or vertual) displacement 虚位移

in term of y 用 y 来表示

incandescent [ˌinkæn'desnt] lamp 白炽灯

in-class education of ideology and politics 课程思政

incombustible [inkəm'bʌstibl] adj. 防火的，不燃的

Incrementalism of Planning Theory 渐进主义规划理论

industrial design 工业设计

industrialization [inˌdʌstriəlai'zeiʃn] n. 工业化

inertia [i'nəːʃjə] n. 惯性，惯量，惰性

infinite ['infinit] adj. 无穷的，无限的；n. 无穷大

infinitesimal [ˌinfini'tesiml] n. adj. 无穷小(的)

infinity [in'finiti] n. 无穷大

inflection [in'flekʃən] n. 弯曲，屈曲

influence line 影响线

information security 信息安全

information superhighway 信息高速公路

Information Technology (IT) 信息技术

infrastructure ['infrəstrʌktʃə(r)] n. 基础设施

ingress and egress 出入口

inkjet printer，bubble jet printer 喷墨打印机

in-place, in-situ 现场的

input device 输入设备

Input/Output(or I/O) Device 输入输出设备

instantaneous velocity 瞬时速度

insufficient disk space 磁盘空间不足

insufficient memory 内存不足

integral ['intigrəl] adj. 整数的，积分的；n.
积分；~ constant 积分常数

integration [inti'greiʃən] n. 积分(法)，求积

intelligent transportation system (ITS)智能交通

intensity [in'tensiti] n. 强度，强烈

interaction diagram 相关图

interchange['intətʃeindʒ] n. 互通式立体交
叉；~ ramp 互通立交匝道

International Standard Organization (ISO) 国际
标准化组织

Internet ['intənet] n. 互联网

Internet Data Center (IDC)互联网数据中心

Internet Explorer IE 浏览器

Internet Service Provider (ISP)互联网服务提
供者，网络服务提供商

Internet telephone 网络电话

interpolation [in,tə:pə'leiʃn] n. 内插法，插
值法

inverse function 反函数

inverse trigonometric function 反三角函数

invited tender 邀请招标

involution [,invə'lu:ʃən] n. 乘方

IP (Internet Protocol)因特网协议

IP address 网际协议地址，网络地址，IP
地址

IR = polyisoprene [,pɔli'aisəupri:n] rubber 聚
异戊二烯橡胶，异戊橡胶

isotropic [,aisəu'trɔpik] adj. 各向同性的

iterative process 迭代过程

jalousie (or louver) window 百叶窗

job-built 现场制作

Johannes Itten [dʒəu'hænis 'itən] n. 约翰内
斯·伊顿(建筑师)

Joost Schmidt [dʒu:st ʃmit] n. 乔斯特·史
密特(建筑师)

Josef Albers ['dʒəusif 'ælbəs] n. 约瑟夫·阿
伯斯(建筑师)

junior beam, secondary beam (or girder)次梁

junk E-mail 垃圾邮件

keyboard ['ki:bɔ:d] n. 键盘

keypad ['ki:pæd] n. (辅助)小键盘

keystone ['ki:stəun] n. 拱心石，拱顶石，
楔石

kick strip, mopboard 踢脚，踢脚板

kilobyte (KB) ['kiləbait] n. [计] 千字节

known [nəun] n. 已知数(物)

L & CM = lime and cement mortar 混合砂浆

Laboratory School of Industrial Design in New
York 纽约工业设计实验学校

lacquerware ['lækəwɛə] n. 漆器

LAN (Local Area Network)局域网

land use plan-making process 用地规划编制
过程

lane [lein] n. 车道；fast ~ 快车道

lane lines 车道线

lap splice 搭接

laptop ['læptɔp，美'læpta:p] n. 笔记本电脑

laser printer 激光打印机

Laszlo Moholy-Nagy ['læzlou 'mɔhəli 'neigi] n.
拉兹洛·莫霍利·纳吉(建筑师)

Law of People's Republic of China on Tenders
and Bids 中华人民共和国招标投标法

lawn [lɔ:n] n. 草坪

layer ['leiə(r)] n. 层(次)，夹层；~ by ~
stack printing 逐层堆叠打印

layman ['leimən] n. 非专业人员，外行

layout ['leiaut] n. 规划，设计，设计图案，
布局

least radius of gyration 最小回转半径

left-turn-only lane 左转专用车道

leg [leg] n. 侧边，直角边

Leibniz ['libniz] n. 莱布尼茨(数学家)

Leonardo da Vinci [liəu'nɑ:dəu də 'vintʃi]
n. 列奥纳多·达·芬奇(意大利画家、
艺术家)

letting ['letiŋ] n. 公开开标

letting people 发包人(party issuing contract)

levee ['levi] n. 防洪堤，码头

lift well 电梯井

lift, electric elevator 电梯

lift-slab construction 升板法施工

light well 采光井

lighting ['laitiŋ] n. 采光；~ and ventilation
采光通风

lightly-reinforced, rare-reinforced, less-rein-
forced 少筋

Appendix 2: Commonly Used English-Chinese Professional Terms for Civil and Architectural Engineering
（附录 2：常用土木建筑类专业词汇英汉对照）

247

limestone［'laimstəun］n. 石灰石

limit［'limit］v. 限制；n.［数］极限

limit state［'limit 'steit］极限状态

linden tree 菩提树，椴树

linear［'liniə］adj. 线性的，直线的；~ algebra 线性代数；~ equation 线性方程；~ function 线性函数；~ programming and benefit-cost models 线性规划和成本利益模型

linear expansion coefficient 线膨胀系数

Lingering Garden 留园

lintel［'lintəl］n.（门窗）过梁，楣

Lions Grove 狮子林

Liquid Crystal Display（LCD）液晶显示器

liquid limit 液限

living room, sitting room, parlor 起居室，客厅

load［ləud］n. 荷载；live ~ 活（荷）载；~-bearing capacity 承载力；~ effect 荷载效应；~-bearing masonry 承重砌体；~-bearing wall 承重墙；~-bearing structure 承载结构

lobby［'ləbi］n. 大厅，休息室

location［lou'keiʃən］n. 定位，放样，位置，地点

log in（login），log on（logon）注册，登录，连网，上网，进入系统

log out（logout）注销，退网，下网，退出系统

logarithm［'ləgəriðəm］n. 对数；~ function 对数函数（=logarithmic function）

logo design 标志设计

longitudinal［ləndʒi'tju:dinl］adj. 纵向的，经线的；~ bar 纵向钢筋

loop［lu:p］n. 环线，环道，回路；v. 使成圈，循环

Louis Kahn［'lu(:)is ka:n］n. 路易斯·卡恩（建筑师）

low（high）-volume 低（高）交通量

low-rise building 低层建筑

Ludwig Mies van der Rohe 路德维希·密斯·凡德罗（德国建筑师）与赖特、勒·柯布西耶、格罗皮乌斯并称四大现代建筑大师。密斯坚持"少就是多（Less is more）"的建筑设计哲学，在处理手法上主张流动空间的新概念。

lump sum 总价合同

Lyonel Feininger［'liənəl 'fainingə］n. 利奥尼·费宁格（建筑师）

macadam［mə'kædəm］n. 碎石路，碎石路面

macro［'mækrəu］n. 宏

mahogany［mə'həgəni］，rosewood，blackwood n. 红木，花梨木，紫檀，玫瑰木

mailbomb［'meilbəm］n. 邮件炸弹

main entrance 主入口

main beam，primary beam（or girder）主梁

mainboard［'meinbɔ:d］（main board）n. 主板

maintainability［mein,teinə'biliti］n. 可维护性

malicious program 恶意程序

manhole［'mænhəul］n. 窨井

manhole cover 窨井盖

Mannheim［'mænhaim］n. 曼海姆（德国城市）

mapping［'mæpiŋ］n. 映射

marble［'ma:bl］n. 大理石，大理岩

Marcel Breuer［'ma:səl 'bru:ə］n. 马塞尔·布鲁尔（建筑师）

Martin Friedman［'ma:tin 'fri:dmən］n. 马丁·弗里德曼（建筑师）

mason［'meis(ə)n］n. 泥瓦匠

masonry［'meisnri］n. 砖石建筑，砌体（工程）；~ structure 砌体结构；~ unit 砌块

mass concrete 大体积混凝土

mass storage technology 海量存储技术

Master of Fishing Nets Garden 苏州网师园

mastic［'mæstik］n. 玛碲脂，胶泥，树脂，嵌缝料

mat foundation 筏式基础，筏基，整体基础

matrix［'meitriks］n. 矩阵

matrix printer 点阵打印机

maximal（or maximum）value 极大值

MAYA software MAYA 软件

mechanical parameter 力学参数

mechanical properties 力学性质

mechanical ventilation 机械通风

mechanics［mi'kæniks］n. 力学，机械学；~ of materials 材料力学

mechanism［'mekənizəm］n. 机理，作用原理

mechanize［'mekənaiz］vt. 使机械化

media player 媒体播放器

median［'mi:diən］n.（公路中间）隔离带，分隔带；~ barrier 中央分隔带护栏，中央路栏

megabyte（GB）['megəbait] n. 兆字节

megahertz（MHz）['megəhəːts] n. 兆赫

megastructure ['megəstrʌktʃə] n. 大型建筑，巨型建筑，巨型结构

membrane ['membrein] n. 膜，薄膜，羊皮纸；~ structure 膜结构；~ curing 薄膜养护

memory ['meməri] n. 存储器，内存

memory bar 内存条

memory card 内存卡

menu ['menjuː] n. 菜单

menu bar 菜单栏

metal window 钢窗

metallic [mi'tælik] adj. 金属的；~ tape 钢卷尺

metamorphic rock 变质岩

method of interpolation 插值法；method of linear interpolation 直线插值法

method of joints 结点法

method of least work 最小功法

method of sections 截面法

microclimate ['maikrəuklaimət] n. 小（微）气候

mid-high-rise building 中高层建筑

mid-rise building 中层建筑

mid-span 跨中；~ moment 跨中弯矩

minimal（or minimum）value 极小值

minimum stirrup ratio 最小配箍率

minor arterial 次干道

minus ['mainəs] prep. 减；adj. 负的

mix（or mixing）proportion（or ratio）配合比

mix design 配合比设计

MlDI（Musical Instrument Digital Interface）音乐设备数字接口，乐器数字接口

mobile broadband modem 移动宽带调制解调器

mobile commerce 移动商务

mobile hard disk，Mobile HD 移动硬盘

Mobile Web（Mobile Internet）移动互联网

modem ['məudem]（即 modulator- demodulator）n. 调制解调器，路由器

moderate-reinforced，proper-reinforced 适筋

modular system 模数制

modular tubes 成束（组合）筒

modulus ['mɔdjuləs] n. 模量，模数

Moholy-Nagy ['mɔhəli 'neigi] n. 莫霍利·纳吉（建筑师）

Mohr's circle of stress 摩尔应力圆

moisture content 含水量，湿度

moment ['məumənt] adj. 片刻的，瞬间的，力矩的；n. 瞬间，弯矩，力矩

moment distribution method 弯矩分配法

moment of inertia 惯性矩

Mona Lisa 蒙娜丽莎（达·芬奇所画之著名人像画）

Monastery of Santa Maria delle Grazie 圣玛利亚（德尔格契）修道院

monitor ['mɔnitə(r)] n. 显示器

monochrome ['mɔnəkrəum; mɔnɔ'krəum] n. 单色，单色画；adj. 单色的，黑白的

monolithic reinforced concrete building 整体式钢筋混凝土房屋

monorail ['mɔnəureil] n. 单轨铁路

monument ['mɔnjumənt] n. 界标，界碑；纪念碑

mortar ['mɔːtə(r)] n. 砂浆；~ bed 砂浆平缝；~ joint 灰缝

mosaic [məu'zeiik] n. 马赛克

mosquito screen，screen window 纱窗

motherboard ['mʌðəbɔːd] n. 主板，母板

Mountain Villa of Greenery 苏州环秀山庄

mouse [maus] n. 鼠标

movement joint，deformation joint 变形缝

mud room（农舍入口的）小屋，泥鞋室

multimedia [mʌlti'miːdiə] n. 多媒体（技术）

multiple strands of wire，steel strand 钢绞线

multiplication [mʌltipli'keiʃən] n. 乘，乘法

multi-processor 多处理器

multistoried（or multistory）building 多层建筑

municipality [mjuːnisi'pæliti] n. 市区，城市

Muslim architecture 伊斯兰教建筑

nanotechnology [nænəutek'nɔlədʒi] n. 纳米技术

National Center for the Performing Arts（中国）国家大剧院

Natural Conservation Zone 自然保护区

natural ventilation 自然通风

necking ['nekiŋ] n. 颈缩

negative number 负数

negotiated tender 协议招标

network administrator 网络管理员

neural network 神经网络

neutral axis 中性轴

Appendix 2: Commonly Used English-Chinese Professional Terms for Civil and Architectural Engineering
（附录 2：常用土木建筑类专业词汇英汉对照）

249

no-fines concrete 无细骨料混凝土

nominal interest rate 名义利率

non-developable surface 不可展曲面

nonferrous metal 有色金属

non-load-bearing infill 非承重填充墙

non-load-bearing wall 非承重墙

nonnegative [nɔn'negətiv] adj. 非负的

nonterminating nonrepeating decimal 无限不循环小数

norm for detailed estimates 预算定额

norm for estimating labor requirements 劳动定额

norm for estimating material requirements 材料定额

norm for preliminary estimates 概算定额

normal ['nɔ:məl] adj. 正交的，法向的；~ distribution curve 正态分布曲线；~ distribution 正态分布；~ stress 法向应力

notebook computer 笔记本电脑

nth-order polynomial n 阶多项式

Num Lock 数字键锁定

numerator ['nju:məreitə(r)] n. 分子

numerical analysis 数值分析

oakum ['əukəm] n. 麻刀，麻丝

object-based system 基于对象的系统

object-oriented system 面向对象系统

obtuse angle 钝角

odometer [əu'dɔmitə] n. 测距仪，里程表

Office Automation (OA) 办公自动化

one-one correspondence 一一对应

one-way slab 单向板

one-way street 单行道，单向交通

only if 仅当

open interval 开区间

open tender 公开招标

open ditch, open trench 明沟

Operating System (OS) 操作系统

optical fiber communication 光纤通信

optimum moisture content 最佳含水量

optimum solution 最优解

ordered pair 有序对

ordinary differential equation 常微分方程

ordinate ['ɔ:dinit] n. 纵坐标

Oriental Pearl TV Tower 东方明珠电视塔

origin ['ɔridʒin] n. 坐标原点

origin-destination (O-D) study (or survey) 起迄点研究，OD 调查

orthogonal (or rectangular) coordinate system 直角坐标系

orthotropic deck 正交各向异性桥面

Oskar Schlemmer ['ɔskə 'ʃlemə] n. 奥斯卡·史雷梅尔(画家和舞台设计师)

osmotic coefficient 渗透系数

osmotic pressure 渗透压力

output device 输出设备

outside finish 外装修

overhanging beam 外伸梁

overhead [,əuvə'hed] n. 企业一般管理费用

overload [əuvə'ləud] n. 超重，超载

over-reinforced 超筋

overtaking 超车；~ lane, passing lane 超车道

overturning moment 倾覆力矩

owner ['əunə] n. 业主

oxide ['ɔksaid] n. 氧化物

pad foundation 独立基础

paint [peint] n. 油漆

painted pottery 彩陶

Palladian architecture, Palladianism 帕拉第奥建筑

Palladio [pə'leidjəu] 帕拉第奥(16 世纪意大利建筑家)

panel ['pæn(ə)l] n. 面板，嵌板

pantry ['pæntri] n. 餐具室，食品室，备餐室

parabola [pə'ræbələ] n. 抛物线

parallel port 并行接口

parallel chord 平行弦

parapet ['pærəpit] wall 女儿墙，护栏，防浪墙

parenthesis [pə'renθisis] n. 圆括号"( )"(= round bracket)

Paris's Louvre museum 巴黎卢浮宫博物馆

parking garage 停车库

parterre [pɑ:'tɛə] n. 花坛，花圃

Parthenon ['pɑ:θinən] n. 帕特农神庙

partial differential equation 偏微分方程

partition [pɑ:'tiʃən] n. 分隔墙；~ screen 隔断；~ wall 隔墙；隔断

password ['pɑ:swə:d] n. 密码，口令

patio ['pætiəu] n. 天井，院子

Paul Klee [pɔ:l kli:] n. 保罗·克利(建筑师)

pavement ['peivmənt] n. 路面，铺装，人行

道；~marking 路面标线

PC (Personal Computer) 个人计算机

pea shingle 绿豆砂，豆砾石

peat [pi:t] n. 泥炭，泥煤，泥炭土

pedestal ['pedistəl] n. 基座，柱脚

pedestrian crossing 人行横道线

pedestrian underpass 人行地道，人行地下通道

pedologist [pi(:)'dɔlədʒist] n. 土壤学家

penstock ['penstɔk] n. 闸门，给水栓，水道，水渠

percentage of steel 配筋率

performance and payment bonds 履约保单和支付保函

performance guarantee 履约保证金

perimeter beam，ring beam 圈梁

perimeter tube and interior core walls 外筒内墙（筒中筒）

permeability [,pə:miə'biliti] n. 渗透性(度)

permeate ['pə:mieit] v. 渗入，渗透

permissible stress，allowable stress 容许应力

perpendicular [,pə:pən'dikjulə(r)] adj. 垂直的，正交的；n. 垂线

perspective [pə'spektiv] n. 透视图

Peter Rothschild ['pi:tə 'rɔθtʃaild] n. 彼得·罗斯柴尔德(景观建筑师)

Phillips Collection in Washington 华盛顿菲利普斯艺术博物馆(收藏馆)

phoney shutter 华而不实的百叶窗，活动遮板

phosphorus ['fɔsfərəs] n. 磷

photogeology [fəutədʒi'ɔl'dʒi] n. 摄影地质

photogrammetry [fəutə'græmitri] n. 摄影测量

Picasso [pi'kæsəu] n. 毕加索(西班牙画家、雕塑家，现代艺术创始人)

pie chart 饼图

pier [piə] n. 码头，桥墩；~foundation 墩式基础

pilaster [pi'læstə(r)] n. 壁柱

pile [pail] n. 桩；~driving 打桩；~foundation 桩基；~groups 群桩

pillared arcade [ɑ:'keid] 带柱拱廊

pilot tunnelling 导洞隧道

pin connection 铰接

pipeline ['paiplain] n. 管线，管道

pipeline gas 管道煤气

pit [pit] n. 试坑

pitch [pitʃ] n. 坡度，高度，间距，节距

pitched roof 坡屋顶

peaceability [pleisə'biliti] n. 和易性

placemaking ['pleis'meikiŋ] n. 场所(空间)营造

placement ['pleismənt] n. 浇捣

plain carbon steel 碳素钢

plain concrete 素混凝土

plan view 平面图

plane analytic geometry 平面解析几何

planning，programming，budgeting system (PPBS) 计划规划预算系统

plaster ['plɑ:stə(r)] n. 石膏，灰泥

plaster of Paris 熟石膏

plastic ['plæstik] adj. 塑性的；~flow 塑流，塑性流动；~limit 塑限；~hinge 塑性铰

plasticity [plæstisiti] n. 塑性

plasticizer ['plæstisaizə] n. 塑化剂

plat roof 平屋顶

plate loading tests (地基的)静载荷试验

plinth [plinθ] (wall) n. 勒脚，底座，基座

plot [plɔt] v. 标桧，绘制，作图；n. 图板，测绘板；~ratio 容积率，用地容积率

plotter ['plɔtə(r)] n. 绘图仪

plug and play (or PnP，P&P) 即插即用

plum blossoms 梅花

plumb [plʌm] n. 垂球，铅垂；adj. 垂直的；~line 铅垂线

plumbing [plʌmiŋ] n. （自来水，卫生）管道

plus [plʌs] prep. 加；adj. 正的；n. 正

plywood ['plaiwud] n. 胶合板

pneumatic caisson 气压沉箱

point bearing pile 端承桩

point of tangency 切点

pointing masonry 清水墙，勾缝砌体

Poisson's ratio 泊松比 μ

polygonal [pə'ligənl] adj. 多边形的

polynomial [,pɔli'nəumjəl] n. 多项式

polypropylene [,pɔli'prəupəli:n] n. 聚丙烯

population density 人口密度

pop-up menu [计] 弹出式菜单

porcelain ['pɔ:slin] n. 瓷，瓷器；adj. 瓷制的

porch [pɔ:tʃ] n. 门廊，走廊

pore water 孔隙水

Portland blast furnace slag cement 矿渣水泥

Appendix 2: Commonly Used English-Chinese Professional Terms for Civil and Architectural Engineering
(附录2：常用土木建筑类专业词汇英汉对照)

251

Portland cement 波特兰水泥，硅酸盐水泥

Portland fly-ash cement 粉煤灰（硅酸盐）水泥

Portland-pozzolana cement 火山灰质硅酸盐水泥

positive number 正数

POST（Power-On Self-Test）加电自检

postanalysis phase 后分析阶段

postmodernism 后现代主义

post-tensioning method 后张法

potassium [pə'tæsiəm] n. 钾

pottery craft 陶瓷工艺

pottery ware 陶器

pounded earth 夯土

pour [pɔ:(r)] v. 浇筑，倾倒

power ['pauə(r)] n. 幂，乘方，势

pozzolana [ˌpɔtsə'la:nə]（volcanic ash）n. 火山灰，水泥与火山灰混合水泥

pozzolana cement 火山灰（质硅酸盐）水泥

preanalysis phase 预分析阶段

pre-built 预制

precast pile 预制桩

precast concrete lintel 预制混凝土过梁

precast reinforced concrete building 预制钢筋混凝土房屋

preliminary [pri'liminəri] adj. 初步的；~ design stage 初步设计阶段；~ estimate 设计概算；~ investigation 初步勘察

pressure ['preʃə] n. 压，压力，压迫，强制，紧迫

pressuremeter test 旁压试验

prestress [pri:'stres] vt. 加预应力；n. 预应力；~ losses 预应力损失

prestressed concrete 预应力混凝土

prestressing force 预应力

pre-tensioning method 先张法

primary beam 主梁

principal arterial 主干道

principle of superposition 叠加原理

principle of virtual displacement 虚位移原理

principle of virtual work 虚功原理

printer ['printə] n. 打印机

printer driver 打印机驱动程序

prism ['prizəm] n. 棱柱体

Pritzker Architecture Prize 普利兹克建筑奖

probability theory and mathematical statistics 概率论与数理统计

product design 产品设计

profile ['prəufail] n. 纵剖面图，断面图，轮廓

program design 程序设计

program evaluation and review technique, PERT 计划评审技术，规划评估与复核法

programming language [计] 编程语言

progress payment 工程进度款

project ['prɔdʒekt] n. 工程，项目

proof by contradiction 反证法

proof by induction [数] 用归纳法证明

prop [prɔp] n. v. 支柱，支撑

property ['prɔpəti] n. 财产，地产，所有权，财产权；n. 性质，特性，性能；~ line 用地红线

property survey 地籍测量，地产测量

proportionality [prəˌpɔ:ʃə'næliti] n. 比例；~ factor 比例因数

proportioning of concrete 混凝土配合比

proposed project 拟建工程

proposed structure 拟建结构

propping ['prɔpiŋ] n. 支撑

props or shores 支撑

protecting cap，safety helmet 安全帽

protecting net 安全网

protective barriers 护栏，防护栏杆

prototype ['prəutətaip] n. 创作原型，样品

public building 公共建筑

public comfort station，public convenience，WC 公共厕所

pulverized coal ash 粉煤灰

pump ['pʌmp] vt. 抽吸，泵送；~ concrete，~ing concrete 泵送混凝土

purlin ['pə:lin] n. 檩条，桁条

pylon ['pailən] n. 桥塔，指示塔，高压线铁塔

quadra ['kwɑ:drə] n. 勒脚

quadrangle ['kwɔdræŋgl]，courtyard house 四合院

quadrant ['kwɔdrənt] n. 象限

quadratic equation 二次方程

quarry ['quɔri] n. 采石场

quasi-permanent ['qweizai-'pə:mənənt] adj. 准永久的

quench [kwentʃ] n. vt. 淬火冷却

quicklime ['kwiklaim] n. 生石灰，氧化钙

radial displacement 径向位移

radial strain 径向应变

radial stress 径向应力

radius ['reidjəs] n. 半径

raft foundation 筏基

rafter ['rɑːftə] n. 椽(子)

railroad flat car 铁路平板货车，铁路平板车

rain-gutter 檐沟，天沟

rain-water leader 水落管

raised or recessed retroreflective pavement markers 凸起或凹入式定向反光路面标线

ramp [ræmp] n. 匝道，坡道，斜坡道；v. 做成斜坡

Random Access Memory (RAM) 随机存储器

range [reindʒ] n. 值域，范围

rate of contraction 收缩率

rate of expansion 膨胀率

ratio of green space, greening rate 绿地率

rationalism ['ræʃənəlizəm] n. 理性主义

Rationalism of Planning Theory 理性主义规划理论

reactive force 反力

Read Only Memory (ROM) 只读存储器

ready-mixed concrete 商品(或预拌)混凝土

real estate 房地产

recess [ri'ses] n. 凹处，深处；vt. 使凹进；~ed veranda 凹阳台

recharge [riː'tʃɑːdʒ] n. 补充(量)；~ well 回灌井

reciprocal [ri'siprəkəl] adj. 倒数的，彼此相反的

reconnaissance [ri'kɔnisns] n. 踏勘，勘测；~ map 踏勘图；~ survey 路线测绘，普查

red cray, adamic earth 红黏土

redundant force 冗余力

refresh rate 刷新率

regional planning 区域规划

registered ['redʒistəd] adj. 注册的；~ architect, licensed architect 注册建筑师；~ planner 注册规划师；~ structural engineer 注册结构工程师

regression [ri'greʃn] n. 回归；~ analysis 回归分析；~ coefficient 回归系数；~ equation 回归方程

regulatory detailed planning 控制性详规

rehabilitation [ˌriːhəˌbili'teiʃn] n. 重建，修复，更新；~ facility 康复设施

reinforce [ˌriːin'fɔːs] vt. 加强，加固，增援，配筋于；~d masonry 配筋砌体

reinforced concrete (or RC, or R. C.) 钢筋混凝土，配筋混凝土

reinforcement [riːin'fɔːsmənt] n. 增援，加强，(混凝土中的)钢筋；~ bar 钢筋；~ stirrup ['stirəp] 箍筋；~ cover 混凝土保护层；~ mat (or mesh) 钢筋网；~ ratio (or percentage) 配筋率

reinforcing work 钢筋工程

reliability [riˌlaiə'biliti] n. 可靠性，可靠度；~ coefficient 可靠性系数

remote sensing 遥感

remote terminal 远程终端

remuneration [riˌmjuːnə'reiʃn] n. 报酬

Rene Descartes ['renei dei'kaːt] n. R. 笛卡尔(法国数学家、哲学家)

resident engineer (施工) 现场工程师

residential [riˌzidenʃ(ə)l] adj. 住宅的，与居住有关的；~ building 居住建筑

residual [ri'zidjuəl] adj. 剩余的，残留的

resiliency [ri'ziliənsi] n. 弹性

resilient [ri'ziliənt] adj. 弹性的，有回弹力的

resistance [ri'zistəns] n. 抵抗力，阻力

resolution [ˌrezə'luːʃn] n. 分辨率

restored building 仿古建筑

resultant [ri'zʌltənt] adj. 合成的；n. 合力

resurface [ˌriː'səːfis] vt. 罩面，重做面层

retaining wall 挡土墙

retarder [ri'tɑːdə] n. 缓凝剂

retarding admixture 缓凝剂

retroreflective [ˌretrəuri'flektiv] adj. 定向反光的，逆向反光的

RGB (Red, Green, Blue) 三原色(红色、绿色、蓝色)

rhomboid ['rɔmbɔid] n. 长菱形

right angle 直角

right of way, ROW 先行权，通行权，路权

right-click 右击

rigid ['ridʒid] adj. 刚性的，坚固的，不易弯曲的；~ foundation 刚性基础

rigid frame and haunch girders 刚性框架-加腋大梁

rise-to-span ratio (rise-span ratio) of an arch

Appendix 2: Commonly Used English-Chinese Professional Terms for Civil and Architectural Engineering
(附录2：常用土木建筑类专业词汇英汉对照)

253

拱的矢跨比(矢高/跨度 = f/l)

rivet ['rivit] n. vt. 铆接，铆钉

riveting，riveted connection 铆接，铆钉连接

road junction 交叉路口，道路交叉点

road kerb 路缘石

road marking 道路标线，路标，道路标记

Robert Hooke ['rɔbət huk] n. R. 胡克(物理学家)

rock mechanics 岩石力学

rock pressure tests 岩石应力试验

rock mass mechanics 岩体力学

rolled [rəuld] adj. 碾压的，轧制的，辊轧的

Romanesque capital 罗马式柱头

roof [ruːf] n. 屋顶，屋盖；~ slab 屋面板；~ garden 屋顶花园；~ live load 屋面活荷载

rotary and percussion 旋转与冲击(钻探)

roundabout ['raundəbaut] n. 环岛，环形交叉口

router ['ruːtə(r)] n. 路由器

row [rəu] n. 行

rubble ['rʌbəl] n. 毛石，碎石；~ wall 毛石墙

rumpus room 娱乐室

running sand 流砂

running speed 行驶速度，运行速度

Russian architecture 俄罗斯建筑

rut [rʌt] n. 车辙；vt. 挖槽于，在…形成车辙

safe mode 安全模式

safety index 可靠指标

safety coefficient 安全系数

safety glass 安全玻璃

sand cushion 砂垫层

satellite city 卫星城市

saturated ['sætʃəreitid] adj. 渗透的，饱和的

saw-tooth ['sɔːˈtuːθ] n. 锯齿形；~ skylight 锯齿形天窗

scaffold ['skæfəuld] n. 脚手架

scale economy 规模经济

scanner ['skænə(r)] n. 扫描仪

schematic representation 图解表示

screed-coat 找平层

screen saver [计] 屏幕保护程序

screw [skruː] n. 螺钉；blunt ~ 圆头螺钉；~ed connection，bolting 栓接，螺栓连接

search engine 搜索引擎

secondary beam 次梁

section view 剖面图

security certificate 安全认证

sedimentary rock 沉积岩

segmental bridge 弓形桥

segregation [ˌsegriˈgeiʃn] n. 离析

seismic ['saizmik] adj. 地震的；~ joint 抗震缝；~ design 抗震设计

self-lifting guy derrick 自升牵索桅杆起重机

self-supporting wall 非承重墙

self-weight 自重；~ structural components 自承重结构构件

Separating of concrete，segregation of concrete 混凝土离析

serial port 串行接口

server ['səːvə(r)] n. 服务器

service area 服务区

serviceability [ˌsəːvisəˈbiliti] n. 适用性

set [set] vi. 凝固

settlement ['setlmənt] n. 沉陷，沉降；~ joint 沉降缝；~ of foundation 地基沉降

sewage ['sjuːidʒ] n. 污水

sewer [sjuːə] n. 下水道；~ system 排水系统

shaft [ʃaːft] n. 矿井，竖井，旋转轴，轴，杆状物

shale [ʃeil] n. [岩]页岩，泥板岩

shallow foundation 浅基础

Shanghai Global Financial Hub 上海环球金融中心

Shanghai Tower 上海中心大厦

shear [ʃiə] n. 剪切；~ stress 剪应力；vane ~ test 十字板剪切试验；~ wall buildings 剪力墙结构房屋；~ wall structure 剪力墙结构；~ wall-frame 框剪；~ wall-haunch girder frame 加腋梁框架-剪力墙；~ strength 抗剪强度

shearing force 剪力

sheathing ['ʃiːðiŋ] n. 护套

shield-driven tunnel 盾构隧道

shore [ʃɔː(r)] n. 斜撑，斜撑柱

shoring ['ʃɔːriŋ] n. 支撑

shortcut key 快捷键

shotcrete ['ʃɔtkriːt] n. 喷射混凝土

shoulder ['ʃəuldə] n. 路肩

shoved mortar joint 挤浆砌筑的灰缝

shrinkage [ˈʃriŋkidʒ] n. 收缩

shuttering [ˈʃʌtəriŋ] n. 模板(结构)

sidewalk [ˈsaidwɔːk] n. 人行道；步行道

sight distance 视距

sign convention 符号约定

signalized intersection 信号(控制)交叉口

significant figures [数] 有效数字

silicon chip 硅片

Silicon Valley 硅谷，加州硅谷

sill [sil] n. 窗台

silt [silt] n. 淤泥，粉粒

silty [silti] adj. 淤泥的，粉质的；~ soil 粉质土

simultaneous equations 联立方程

simple function 简单函数

simply supported beam 简支梁

sine curve 正弦曲线

single door 单扇门

single reinforcement 单筋

single-family split-level house 独户错层式住宅

singly reinforced beam 单筋梁

Sinosteel International Plaza 中钢国际广场

site clearing 清(理)场(地)

site investigation 场地勘察

skybridge [ˈskaibridʒ] n. 天桥

skylight [ˈskailait] n. 天窗

skyline [ˈskailain] n. 天际线

slab [slæb] n. 平板；roof ~ 屋面板；floor ~ 楼板

slaked lime 熟石灰，消石灰，氢氧化钙

slenderness [ˈslendənis] n. 长细比

sliding door 推拉门

sliding window 水平推拉窗

slip [slip] v. n. 滑移；~ road 匝道

slipform construction，slipforming 滑模施工

slope [sloup] n. 斜率，坡度，倾斜(角)，斜坡

sluice [sluːs] n. 水闸，水门

slump [slʌmp] n. 坍落度

snow load 雪(荷)载

social impact assessment 社会影响评价

social networking website 社交网站

social sustainability 社会可持续性

socio physical space 社会物质(物理)空间

soft under-lying layer，soft substratum 软弱下卧层

software engineering 软件工程

softwood [ˈsɔftwud] n. 软木

soil [sɔil] n. 土壤，土地；~ mechanics 土力学；~ mechanics survey 地基勘察；~ nailing 土钉支护，土钉(墙)

soil-sampling 取土样

solid [ˈsɔlid] n. 固体；adj. 固体的，坚固的

solidus [ˈsɔlidəs] n. 斜线分隔符，即"/"

solution 解法

sound box 音箱

sound card 声卡

source code 源代码

Southern California School of Design 南加州设计学校

spacer bar 架立筋，定位钢筋

spacing [ˈspeisiŋ] n. 间距

spade [speid] n. 铲，铁锹

spam [spæm] n. 垃圾邮件

span [spæn] n. 跨度(距)，孔距

spandrel [ˈspændrəl] n. 托梁；~ beams 外墙托梁

spatial [ˈspeifəl] adj. 空间的，立体的

specific gravity 比重

sphere [sfiə] n. 球，球体，球面

spherical [ˈsferikəl] adj. 球形的，球面的

spheroidal [sfiəˈrɔidəl] adj. 椭球体的，球状的

spillway [ˈspilwei] n. 溢洪道，泄洪道

spiral [ˈspaiərəl] n. adj. 螺旋(的)；~ stairs，~ staircase 螺旋楼梯

Spoonbridge and Cherry 匙桥与樱桃(雕塑)

spray [sprei] vt. 喷射；~ er 喷雾器，洒水车

spreadsheet [ˈspredʃiːt] n. 电子表格

spring constant 劲度系数(弹簧弹性系数)

spruce tree 云杉树

spyware [ˈspaiwɛə(r)] n. 间谍软件

sq ft(=ft^2)平方英尺

square bracket 方括号"[ ]"

square root 平方根

stability [stəˈbiliti] n. 稳定性

stack bond (stack-bond)横竖通缝砌法，对缝砌法

staggered [ˈstægəd] adj. 交错的，错列的

stair [stɛə(r)] n. 楼梯；~ clearance (or headroom)楼梯净空高度；~ rail 楼梯栏杆

Appendix 2: Commonly Used English-Chinese Professional Terms for Civil and Architectural Engineering
（附录 2：常用土木建筑类专业词汇英汉对照）

255

（或扶手）； ~ step 楼梯踏步； ~ string（or stringer）楼梯梁； ~ case 楼梯间； ~ well 楼梯井

stakeholder［'steikhəuldə(r)］n. 股东

standard brick 标准砖

standard deviation 标准差

standard penetration test（SPT）标准贯入度试验

state electricity 静电

static cone penetration test 静力触探； static load test of pile 单桩静荷载试验

statically determinate 静定的； ~ truss 静定桁架

statically indeterminate 超静定的，静不定的； ~ truss 超静定桁架

statics［'steitiks］n. 静力学，静止状态，静态

steel structure 钢结构

steel forms 钢模板

steepness［'sti:pnis］n. 倾斜度

step［step］n. 踏步

stiffness or displacement method 刚度法，位移法

stiffness，rigidity 刚度

stirrup［'stirəp］n. 箍筋； ~ spacing 箍筋间距

stope 采场

store room 贮藏室

storey height 层高

straight angle 平角（一条射线绕它的端点旋转，当始边和终边在同一条直线上，方向相反时所构成的角）

strain［strein］n. 应变

strength［strenθ］n. 强度

stress［stres］n. 应力； ~ distribution 应力分布； ~ concentration 应力集中； ~ reversals 应力反复； ~ -strain diagram 应力-应变图； ~ -strain relationship 应力-应变关系； ~ -strain curve 应力-应变曲线

stretcher unit 顺砌块材

stringer［'stringə(r)］n. 桁条，纵向轨枕

strip foundation 条形基础

structural［'strʌktʃərəl］adj. 结构的； ~（or structure）mechanics 结构力学； ~ design 结构设计； ~ engineer 结构工程师； ~ reliability theory 结构可靠性理论； ~ draw-

ings 结构图

structure［'strʌktʃə(r)］n. 结构，构造，建筑物

subcontractor 分包商

subgrade［'sʌbgreid］n. 路基，地基

submenu［'sʌbmenju:］n. ［计］子菜单

substandard［ˌsʌb'stændəd］adj. 标准以下的，不合规格的；n. 低标准

substrate［səb'streit］n. 地层，基质

subtraction［səb'trækʃən］n. 减，减法

sun louver 遮阳板

sunshine spacing 日照间距

super high-rise building 超高层建筑

superelevation［ˌsju:pəˌeli'veiʃən］n. 超高

superintendent［ˌsju:pərin'tendənt］n. 管理员，指挥人，总段长

superposition［ˌsju:pəpə'ziʃən］n. 迭加，重迭

supervisor［'sju:pəvaizə(r)］n. 监管员

supplementary［ˌsʌpli'mentəri］adj.（互）补的

support［sə'pɔ:t］n. v. 支座，支持； ~ reaction 支座反力； ~ ing block 支座； ~ ing layer 持力层

surcharge［'sə:tʃɑ:dʒ］n. v. 超载，过载

surety［'ʃuərəti］n. 担保人

surrounding rock 围岩

surroundings［sə'raundiŋz］n. 环境

survey［sə:'vei］n. vt. 测量，调查，勘定

surveyor［sə'veiə］n. 测量员

susceptibility［səˌsepti'bility］n. 磁化率

suspended-cable structure 悬索结构

suspender［sə'spendə(r)］n. 吊杆； ~ cable 吊索

sustainable development 可持续发展

sustainable spatial planning 可持续空间规划

Swallows Garden（Yan Garden）常熟燕园

sway［swei］v. 晃动，侧移

Sydney Opera House 悉尼歌剧院

symmetry［'simətri］n. 对称(性)

system of equations 方程组

tangent［'tændʒənt］n. ［数］正切；adj. 正切的； ~ line 切线

tarlike［'ta:laik］adj. 像焦油的，焦油状的

taskbar［'ta:skba:］n. ［计］任务栏

TBM（Tunnel Boring Machine）隧道掘进机

technical design stage 技术设计阶段

tee［ti：］n. 三通管

tees, channels T 型钢，槽钢

teleconference［'telikɔnfrəns］n.［通信］远程会议，电话会议

temper［'tempə］n. （钢等）回火；vt. vi. （使）回火

temperature joint 伸缩缝

tempered glass, reinforced glass 钢化玻璃

template file 模板文件

tenant［'tenənt］n. 承租人，房客；vt. 出租

tenderee［'tendə：ri］n. 招标人，招标方

tendon［'tendən］n. 筋，索

tensile［'tensail］adj. 拉伸的；~ strength 抗拉强度；~ reinforcement, tension reinforcement 受拉钢筋

tension［'tenʃ(ə)n］n. 拉力，张力，拉伸，拉紧

terra cotta 琉璃瓦

terrace［'terəs］n. 露台；~ roof 平屋顶

terrazzo［te'rætsəu］n. 水磨石

Terzaghi bearing capacity theory 太沙基承载力理论

Terzaghi consolidation theory 太沙基固结理论

the absolute-value function 绝对值函数

The Elements of Architecture《建筑学要素》

the first derivative 一阶导数

the identity function 恒等函数

The Internet of Things 物联网

the second derivative 二阶导数

The Ten Books on Architecture《建筑十书》

the triangle inequality 三角不等式（即：在任何三角形中，任意两边之和大于第三边）

the unit distance 单位长度

theodolite［θi'ɔdəlait］n. 经纬仪

theorem［'θiərəm］n. 定理，法则

Theorem of Pythagoras［pai'θægərəs］毕达哥拉斯定理，勾股定理

theoretical mechanics 理论力学

Theory of Swelling-Rising Concrete Arch with Tree-Root-Shaped Cable Bolt Reinforcement 胀锚拱理论

thermal［'θə：məl］adj. 热的；~ insulation 隔热；~ insulation wall 保温隔热墙体

thermal conductivity 热传导率

thermoplastic［,θə：məu'plæstik］adj. 热塑（性）的；n. 热熔物，热塑

thin-wall webbed I-beam 薄腹板工字梁

thoroughfare［'θʌrəfɛə(r)］n. 大道，要道

thread［θred］n. 螺纹；~ steel bar, deformed bar 螺纹钢筋

Three Treasures of Chinese arts and crafts：Fuzhou bodiless lacquerware, Beijing cloisonné and Jingdezhen porcelain 中国传统工艺"三宝"：福州脱胎漆器、北京景泰蓝和江西景德镇瓷器

three-dimensional［'θri：di'menʃənl］adj. 立体的，三维的，三度空间的

through ventilation 穿堂风

thumbnail［'θʌmneil］n. 缩略图

tie［tai］n. 连系（筋），拉杆；v. 绑扎；~ bar 拉杆；~ plate, stay plate 连接板，缀板

tile［tail］n. 瓦片，瓷砖；hollow ~ 空心砖

timber［'timbə］n. 木材，木料；~ cruise 森林勘察；~ floor 木地板；~ pile 木桩；~（or wood）structure 木结构

title bar［计］标题栏

toilet［'tɔilit］n. 盥洗间，浴室，厕所，便池

token［'təukən］n. 令牌；dynamic ~ 动态口令牌

top chord 上弦杆

topographic［tɔpə'græfik］adj. 地形（测量）的

topology［tə'pɔlədʒi］n. 拓扑学，拓扑关系

torque［tɔ：k］n. 扭矩，扭转

torsion［'tɔ：ʃən］n. 扭转，扭矩

total and differential settlement 总沉降量和沉降差

tower crane 塔式起重机

Townscape Movement 城镇景观运动

tracing paper 描图纸

traffic［'træfik］n. 交通；~ accident 交通故事；~ capacity analysis 通行能力分析；~ congestion 交通拥堵；~ engineering 交通工程；~ flow analysis 交通流（量）分析；~ jam 交通堵塞；~ regulation 交通规则；~ sign 交通标志；~ signal［交］交通信号，红绿灯；~ volume survey 交通量调查

transition［træn'ziʃn, træn'siʃn］n. 转变，过渡；~ curve［建］缓和曲线，［数］过渡曲线

transom［'trænsəm］n. 门上的亮子

transportation planning 交通规划

Appendix 2: Commonly Used English-Chinese Professional Terms for Civil and Architectural Engineering
（附录2：常用土木建筑类专业词汇英汉对照）

257

transverse［ˈtrænzvəːs］adj. 横向的，横断的

transversal surface 横截面

trap［træp］n. 存水弯

travel lane 行车道

treatment of elevation 立面处理

trench［ˈtrentʃ］n. 沟，排水沟，探沟，探槽

triangle［ˈtraiæŋgl］n. 三角形；right ~ 直角三角形

triangulation［traiˌæŋgjuˈleiʃən］n. 三角测量

triaxial［traiˈæksiəl］adj. 三（维）轴的，空间的；~ compression test 三轴压缩试验

trigonometric function 三角函数

trip distribution 出行分配

trip generation 出行生成量

trough［trɔf］n. 槽，水槽，饲料槽，海槽

truck crane 汽车吊，汽车起重机

truss［trʌs］n. 桁架；~ structure 桁架结构

tube structure, hull core structure 筒体结构

tube-shaped apartments 筒子楼

tubular［ˈtjuːbjulə］adj. 管的；~ steel scaffolding 钢管脚手架

tungsten［ˈtʌŋstən］n. 钨

tunnel［ˈtʌn(ə)l］n. 隧道，地道

Turing Test 图灵试验

turnbuckle［ˈtəːnˌbʌkl］n. 花篮螺丝，套筒螺母

turnpike［ˈtəːnpaik］n.［税］收费高速公路

twist［twist］vt. 扭，拧，使扭转；n. 扭曲

Twitter［ˈtwitə(r)］n. 推特，微博

two-way slab 双向板

two-way reinforcement 双向配筋

two-way slabs of reinforced concrete 钢筋混凝土双向板

type-form 形制，造型，型制

U. S. Geological Survey 美国地质测量局

U. S. National Geodetic Survey 美国国家大地测量局

undisturbed samples 原状样

unequal-leg angle 不等边角钢

Uniform Resource Locator（URL）统一资源定位器

uniformly distributed load 均布荷载

United Nation Climate Panel 联合国气候小组

unit-load method 单位荷载法

Universal Serial Bus（USB）通用串行总线

unknown［ˌʌnˈnəun］n. 未知数；~ quantity 未知量

unnotched bar, plain bar 光面钢筋

unreinforced masonry 非配筋砌体

upgrade［ˈʌpgreid］v. n. 升级，更新

upload［ˌʌpˈləud］v. 上传，上载，加载

UPS（Uninterruptible Power Supply）不间断电源

urban［ˈəːbən］adj. 城市的，都市的；~ amenity 城市便利设施；~ and rural（or ~ -rural）planning 城乡规划；~ base map 城市底图；~ comprehensive planning 城市总体规划过程；~ design 城市设计；~ ecology 城市生态学；~ economics 城市经济学；~ ecosystem 城市生态系统；~ geography 城市地理学；~ land use planning 城市土地利用规划；~ master（or comprehensive, or general, or overall）plan 城市总体规划；~ regeneration 城市更新；~ sociology 城市社会学

Urban Planning Society of China 中国城市规划学会

Urban Transportation Model System（UTMS）城市交通模型系统

urbanist［ˈəːrbənist］n. 城市学家

urbanization［ˌəːbənaiˈzeiʃn］n. 城市化，都市化

urinal［juəˈrainl, ˈjuərinl］n. 小便斗，小便槽

USB flash disk U 盘

user account 用户账号

utility［juːˈtiliti］n. 公用设施，公用事业，效用；~ room 杂物间

Utopianism of Planning Theory 规划理论的空想主义（乌托邦主义）

vanadium［vəˈneidiəm］n.［化学］钒

variable［ˈvɛəriəbl］adj. 变化的；n. 变量

vector［ˈvektə］n. 向量，矢量

veneer［vəˈniə］n. 饰面，镶板，饰面砖，墙面砖

ventilating skylight 通风天窗

ventilation［ˌventiˈleiʃn］n. 通风；~ shaft 通风井

vernacular［vəˈnækjulə］n. 民居，乡村民宅，民间风格；~ architecture 乡土建筑

vertical［ˈvəːtikəl］adj. 垂直的，直立的；~ progression 竖向拓展，竖向系列

VGA（Video Graphics Adapter）视频图形适配器

viaduct ['vaiədʌkt] n. 高架桥

video conferencing 视频会议，电视会议

villa [vilə] n. 别墅

Virtual Private Network（VPN）虚拟专用网

Virtual Reality（VR）虚拟现实

visualization [ˌviʒuəlaiˈzeiʃn] n. 可视化

vitreous ['vitriəs] adj. 玻璃的，玻璃状（质）的

volcanic [vɔlˈkænik] adj. 火山的；~ ash 火山灰

waffle ['wɔfl] n. 华夫格子松饼，华夫饼筒

wall between two windows 窗间墙

WAP（Wireless Application Protocol）无线应用协议

Warring States Times（or Period）战国时期

Wassily Kandinsky ['wɔsili 'kændinski] n. 瓦西里·康定斯基（建筑师）

Water Cube, The National Swimming Center 水立方，国家游泳中心

water distribution system 配水系统

water lily 睡莲，百合，荷花

water supply and drainage 给水排水

water treatment 水处理

water vegetation 水草

water-cement ratio（w/c）水灰比

watercolor painting 水彩

waterproof barrier, aquitard 防水层

water-reducing agent 减水剂

watershed ['wɔːtəʃed] n. 分水岭，流域，汇水区

waterside pavilion 水榭

watertight concrete 防水混凝土

wavelet ['weivlət] n. 小波

wayfinding signage ['sainidŋ] 寻路标牌

weak concrete 低标号混凝土，低强度混凝土

weathering ['weðəriŋ] n. 风化

weathering steel 耐候钢

web [web] n. 腹板，梁腹

wedge [wedʒ] n. 楔子，楔形物

weigh station 称重站，地磅站

weigh-in-motion（WIM）system 动态称重系统

weighted means 加权平均值

weighting coefficients 加权系数

Weimar ['waimɑː] n. 魏玛市（德国城市）

weld [weld] n. v. 焊接，熔接，焊缝，焊点

weldability [weldəˈbiliti] n. 可焊性

welding ['weldiŋ] n. 焊接，焊缝连接

white cement 白水泥

widely spaced perimeter tube 大柱距外框筒

William Morris ['wiliəm 'mɔris] n. 威廉·莫里斯（英国设计师、诗人、早期社会主义活动家）

wind direction diagram（wind rose）风向玫瑰图

wind load 风荷载

window blind, sun blind 窗帘

Windows Operating System, Windows 操作系统

wireless ['waiələs], cordless ['kɔːdləs], tetherless ['teðə(r)ləs] adj. 无线的，无绳的

wireless network adaptor 无线网络适配器

without bound 无界，无限

WLAN（Wireless Local Area Network）无线局域网

Word Processing System（WPS）文字处理系统

work period, period of agreement 工期

workability [ˌwəːkəˈbiləti] n. 和易性，可操作性

working（or construction）drawing design stage（or phase）施工图设计阶段

workload ['wəːkləud] n. 工作量

World Commission on Environment and Development 世界环境与发展委员会

worm's eye view 仰视图

WWW（World Wide Web）万维网

wythe [wiθ] n. （建筑中）砖的厚度，一皮砖

x-direction x 方向

y-axis y 轴

y-intercept y 截距

Yangtze River Delta 长江三角洲

yield [jiːld] n. 屈服；vi.（~ to）屈服于，屈从；~ point 屈服点；~ strength 屈服强度；~ stress 屈服应力

Young's modulus, elastic modulus 杨氏模量，弹性模量 E

zebra road markings 斑马线

zero emission building 零排放建筑

References

（参考文献）

[1] 吴炯圻. 数学专业英语[M]. 3 版. 北京：高等教育出版社, 2019.

[2] 卜艳萍, 周伟. 计算机专业英语[M]. 3 版. 北京：人民邮电出版社, 2017.

[3] 张燕, 沈奇, 付弘, 等. 计算机专业英语[M]. 2 版. 西安：西安电子科技大学出版社, 2004.

[4] 孟庆元. 力学专业英语[M]. 哈尔滨：哈尔滨工业大学出版社, 2002.

[5] 宋云连, 崔亚楠. 道路桥梁与交通工程专业英语[M]. 北京：中国水利水电出版社, 2012.

[6] 苏小卒. 土木工程专业英语(上册)[M]. 上海：同济大学出版社, 2000.

[7] 苏小卒. 土木工程专业英语(下册)[M]. 上海：同济大学出版社, 2003.

[8] 惠宽堂, 王泽军, 姚仰平, 等. 土木工程英语[M]. 北京：中国建材工业出版社, 2003.

[9] 邓贤贵. 建筑工程英语[M]. 2 版. 武汉：华中理工大学出版社, 1997.

[10]《建筑工程专业英语教程》编写组. 建筑工程专业英语教程[M]. 武汉：武汉工业大学出版社, 1995.

[11] J. F. Young, S. Mindess, R. J. Gray, et al. 土木工程材料科学与技术(英文版)[M]. 北京：中国建筑工业出版社, 2006.

[12] Z. Y. Shen. Introduction of Civil Engineering(土木工程概论)[M]. Beijing：China Architecture & Building Press, 2005.

[13] 赵纪军, 陈晓彤. 城乡规划专业英语[M]. 武汉：华中科技大学出版社, 2013.

[14] 蒋山, 应宜文. 建筑学专业英语[M]. 2 版. 北京：中国建筑工业出版社, 2013.

[15] 王一, 岑伟. 建筑学专业英语[M]. 北京：中国建筑工业出版社, 2009.

[16] 姜海燕. 建筑专业英语[M]. 北京：中国建材工业出版社, 2003.

[17] Team SCUT-POLITO '长屋计划' 解析[EB/OL]. 个人图书馆, (2018-08-14)[2022-04-10]. http：//www. 360doc. com/content/18/0814/04/32324834_778081530. shtml.

[18] 捌抖视频, 新浪网. 12 个世界上著名的旅游景点, 令人失望的旅游 "陷阱", 还想去吗？[EB/OL]. (2022-03-11)[2022-04-06]. http：//k. sina. com. cn/article_5154305419_13338758b00101abm7. html.

[19] 徐卫国团队. 机器人 3D 打印混凝土书屋[Z/OL]. (2021-03-29)[2022-04-06]. 清华大学建筑学院微信公众号(SATsinghua)新闻.

[20] 徐卫国教授跨学科团队. 机器人 3D 打印混凝土建筑——武家庄一农户喜入新居[Z/OL]. (2021-09-09)[2022-04-06]. 清华大学建筑学院微信公众号(SATsinghua)前沿.

[21] L. Taylor. Urbanized Society[M]. Santa Monica, California：Goodyear Pub. Co, 1980.

[22] S. Hanson. The Geography of Urban Transportation[M]. New York：Guilford Press, 1986.

[23] P. R. Berke, D. R. Godschalk, E. J. Kaiser. Urban Land Use Planning[M]. 5th ed. Urbana：University of Illinois Press, 2006.

[24] P. Naess. Urban Planning and Sustainable Development[J]. European Planning Studies. 2001, 9 (4)：503-524.

[25] E. P. Popov. Mechanics of Materials[M]. 2nd ed. London：Prentice/Hall International, Inc., 1978.

[26] J. J. Waddell. Concrete Construction Handbook[M]. 2nd ed. New York：McGraw-Hill Book Company, 1974.

[27] N. S. Steinhardt. Chinese Imperial City Planning[M]. Honolulu：University of Hawaii Press, 1990.

［28］A. J. Catanese，J，C. Snyder. Urban Planning［M］. New York：McGraw-Hill，1988.

［29］占丰林. 高等教育教学改革之课程思政探索——以《建筑经济与管理》课程为例［J］. 四川建筑，2021，41(3)：278-280.

［30］占丰林. 高温环境下岩体锚固机理——"胀锚拱理论"［J］. 辽宁工程技术大学学报(自然科学版)，2008，27(5)：686-688.

［31］F. L. Zhan，P. Ye. ANSYS Simulating Analysis of 3D Single-Cable Temperature Model of Fully Grouted Cable Bolts［C］// Proceedings of 2012 International Conference on Electric Technology and Civil Engineering(ICETCE 2012). USA：IEEE Computer Society ，2012(3)：1391-1394.

［32］F. L. Zhan，P. Ye. ANSYS Simulating Analysis of a 3D High-Temperature Stope Reinforced in Advance with Fully Grouted Cable Bolts［C］// Applied Mechanics and Materials，Progress in Civil Engineering，Part 1. Proceedings of the 2nd International Conference on Civil Engineering，Architecture and Building Materials(CEABM 2012).

［33］F. L. Zhan，P. Ye. Construction Techniques and Mechanism of Pre-Anchoring Fissured Stope Hangingwall by Fully-Grouted Cable Bolts［C］// Applied Mechanics and Materials，Advances in Civil and Industrial Engineering Ⅳ，Part 1. Proceedings of the 4th International Conference on Civil Engineering，Architecture and Building Materials(CEABM 2014).